"A new series that promises to blend scholarly research with popular appeal.... comprehensive, authoritative.... excellent references."

—Clarence Petersen,
Chicago Tribune

Bantam/Britannica Books

Unique, authoritative guides to acquiring human knowledge

What motivates people and nations? What makes things work? What laws and history lie behind the strivings and conflicts of contemporary man?

One of mankind's greatest natural endowments is the urge to learn. Bantam/Britannica books were created to help make that goal a reality. Distilled and edited from the vast Britannica files, these compact introductory volumes offer uniquely accessible summaries of human knowledge. Technology and science, politics, natural disasters, world events—just about everything that the inquisitive person wants to know about is fully explained and explored.

BANTAM/BRITANNICA BOOKS

Energy

The Fuel of Life

Prepared by
the Editors of
Encyclopaedia
Britannica

Bantam edition/August 1979

For information address: Bantam Books, Inc.
ISBN 0-553-13105-2
Published simultaneously in the United States and Canada

Bantam Books are published by Bantam Books, Inc. Its trademark, consisting of the words "Bantam Books" and the portrayal of a bantam, is registered in the United States Patent Office and in other countries. Marca Registrada.
Bantam Books, Inc.,
666 Fifth Avenue, New York, New York 10019

Printed in the United States of America
0 9 8 7 6 5 4 3 2 1

Foreword: Knowledge for Today's World

One of mankind's greatest natural endowments is the urge to learn. Whether we call it knowledge-seeking, intellectual curiosity, or plain nosiness, most people feel a need to get behind the newspaper page or the TV newscast and seek out the background events: What motivates people and nations? What makes things work? How is science explained? What laws and history lie behind the strivings and conflicts of contemporary man? Yet the very richness of information that bombards us daily often makes it hard to acquire such knowledge, given with authority, about the forces and factors influencing our lives.

The editors at Britannica have spent a great deal of time, over the years, pondering this problem. Their ultimate answer, the 15th Edition of the *Encyclopaedia Britannica*, has been lauded not merely as a vast, comprehensive collection of information but also as a unique, informed summary of human knowledge in an orderly and innovative form. Besides this work, they have also thought to produce a series of compact introductory volumes providing essential information about a wide variety of peoples and problems, cultures, crafts, and disciplines. Hence the birth of these Bantam/Britannica books.

The Bantam/Britannica books, prepared under the guidance of the Britannica's Board of Editors, have been distilled and edited from the vast repository of information in the Britannica archives. The editors have also used the mine of material in the 14th Edition, a great work in its own right, which is no longer being published because much of its material did not fit the design imposed by the 15th. In addition to these sources, current Britannica files and reports—including those for annual yearbooks and for publications in other languages—were made available for this new series.

All of the Bantam/Britannica books are prepared by Britannica editors in our Chicago headquarters with the assistance of specialized subject editors for some volumes. The Bantam/Britannica books cover the widest possible range of topics. They are current and contemporary as well as cultural and historical. They are designed to provide *knowledge for today*—for students anxious to grasp the essentials of a subject, for concerned citizens who want to know more about how their world works, for the intellectually curious who like good reading in concise form. They are a stepping stone to the thirty-volume *Encyclopaedia Britannica*, not a substitute for it. That is why references to the 15th Edition, also known as *Britannica 3* because of its three distinct parts, are included in the bibliographies. While additional research is always recommended, these books are complete unto themselves. Just about everything that the inquisitive person needs to catch up on a subject is contained within their pages. They make good companions, as well as good teachers. Read them.

The Editors,
Encyclopaedia Britannica

Contents

Introduction: The Earth and Today's Energy Crisis

Everyone knows what energy is: it is something of which a shortage produces a crisis. An energy crisis means lines of cars at gas stations. It means schools closed in winter for lack of fuel oil or, in summer, office workers in skyscrapers stifling for lack of air conditioning. It means grounded jet liners and factories grinding to a halt. It means food production threatened by shortages of fertilizers, pesticides, and herbicides and by the failure of pumps for irrigation and of trucking to get produce to market.

On the level of power politics an energy crisis means the Organization of Petroleum Exporting Countries (OPEC). Because three-fifths of the world's known oil reserves are in the Middle East, the oil companies supplying the industrialized countries have invested heavily in developments there. Production was originally cheap, and the "oil sheikhs" were accommodating. The 6% of the world's population that lives in the United States and consumes 30% of the world's energy long ago exceeded its domestic oil production capacity but for many years could burn imported oil at bargain prices. In 1973, however, the Arab states, which dominated OPEC, took matters into their own hands. In the Yom Kippur war, when Egypt and Syria attacked Israel on October 6, the other Arab states used oil as a means of intervention. They imposed an embargo on oil supplies to countries helping Israel. On October 19 they stopped oil supplies to the United States. Although there was a cease-fire in November, the embargo was not lifted until March; oil flowed again—but at a price determined thenceforward by OPEC.

Oil has also become a weapon in civil war. In the 1978 insurrection against the shah of Iran, a strike of oil workers brought about a complete stoppage of production and refining. In the country that had supplied 12% of the oil requirements of Western industrialized countries, there was not enough oil to meet even domestic requirements.

The Iranian uprising coincided with a decision by OPEC to impose a phased increase during 1979 of 15% in the price of crude oil, which would mean an additional cost of about $10 billion to the United States. This was at a time when the American dollar, which had at one time been the world's hardest currency, was in serious difficulties because of a balance of payments deficit. In December 1978 Mexico fixed the price of its crude oil at $14.18 per barrel, approximately nine times the price the United States had been paying for Middle Eastern oil in the 1960s.

The industrial countries have become hostage to oil, the fluid energy that powers motorcars, diesel engines, ships, jet aircraft, and electric generating stations, and also, as a chemical repository, serves as the source of plastics, man-made fibers, dyestuffs, drugs, fertilizers, fungicides, pesticides, and herbicides.

Nature's investment in a cupful of oil is millions of microscopic plants and animals and hundreds of millions of years in time. The fuel oil in such a cupful might heat a house for fifteen minutes or push a supertanker for a little longer than a second through perhaps three feet of ocean. As kerosene it could fly a jumbo jet about twenty feet in one-thirtieth of a second. In the form of gasoline, it could power a small car for nearly a mile in about a minute.

Since most of the Earth's 300-million-year accumulation of oil has been used up in the past fifty years, the oil economy is on a one-way trip that is rapidly running out of gas. Likewise, the Earth's supply of natural gas is just as finite as that of oil, although the reserves have not been as accurately estimated. Projections of when the world's natural gas will run out vary widely, but some experts believe that the supply has already passed its peak.

Coal, the third fossil fuel, has been conserved by default in recent decades because of the wider application and greater convenience of oil and gas, and it will last considerably longer. The coal supply, used as solid fuel, could last another four hundred years. It is, of course, possible to gasify or liquefy coal. Coal could even be gasified in the bowels of the Earth. At a high cost it can be converted into fuel oil by hydrogenation, as was done extensively in Germany to keep the war machines running in the last year of World War II. But conservation of coal (or of any fuel) does not just mean not using it but also means using it more efficiently. The conventional uses of coal were, until recent times, notoriously extravagant. Soot-blackened industrial areas are testimony to the inefficient combustion of coal. Power engineers have greatly improved the ways in which coal is burned to produce power. Modern methods promise even greater efficiency. Applying the principles of plasma physics, magnetohydrodynamics

(MHD) can even generate electricity from the flame itself.

In the meantime, the worldwide demand for energy continues to climb, doubling every decade, which makes the development of alternative energy sources imperative. In 1961 the United Nations (UN) called a Conference on New Sources of Energy. It was concerned with the nations of the Third World and how they, too, might acquire industrial prosperity. The advanced countries were not unduly worried. They realized that oil supplies, being an unrenewable resource, would give out eventually, but meanwhile they had abundant supplies at sixteen cents a barrel from the Middle East, with no premonition of OPEC. And there was nuclear energy. The Third World countries could not really afford to be dependent on oil. In the absence of reasonably priced small-scale nuclear reactors, which had not materialized, nuclear energy was very much a case of "unto everyone that hath shall be given"; the costly megawatt nuclear stations were economically viable only where grid distribution or established industrial complexes existed.

The irony, amounting to cynicism, of the 1961 intergovernmental conference was that the prestigious experts, commissioned by the governments of the highly industrialized countries, were wagging their fingers and enjoining the delegates from the Third World to develop thrifty alternative sources of energy. But these were the sources that the industrialized countries were extravagantly ignoring and were to continue to ignore (or, at best, to flirt with) until the OPEC-induced crisis twelve years later. Even the title of the conference was ironic because the "new" sources were, in fact, primordial—the sun, the winds, the tides, running water, bioenergy, and geothermal energy, the heat from the Earth's crust.

Nuclear energy, the newest source of all, was scarcely mentioned on the pretext that it had been dealt with at the UN conferences on the Peaceful Uses of Atomic Energy in 1955 and 1958. In fact it would have been tactless to remind the Third World of hopes deferred. There were already misgivings about the availability of uranium for the grandiose schemes for nuclear fission reactors, which the industrialized countries were planning as their standoff for the energy gap that would arise from the exponential increase in demand and the foreseeable (but not then imminent) depletion of the world's oil reserves. Public concern about a "plutonium economy" and the intractable problem of the disposal of fission waste products was beginning to rumble. The siting of nuclear power plants was becoming problematical when "environmental impact" was being imposed on the cost-benefit and lead-time calculations.

Fusion energy—that *deus ex machina* solution to the nuclear dilemma, with no waste problems and as much fuel as there is deuterium in the seven seas—was not progressing as well as had been expected in the first, fine, careless rapture of 1955. In terms of time it did not fit into the calendar. So, in Rome in 1961, we had the moment of truth not only for the Third World but also for the highly developed countries who might have listened to their own lecturers and have vigorously promoted research and development of alternative sources of energy. But oil was still cheap, and there was the glamor of the space race to engage the imagination.

In the immensity of the universe, the Earth's local generating station is the sun. The sun's energy is contributed not only in terms of light and radiant heat, converted by photosynthesis into food calories for living creatures and into the coal, oil, and natural gas of the geological deposits, but also as the driving force of that engine, the globe itself. It activates the climate, the air currents, the ocean currents, the winds, the rain, and the flowing rivers. When we talk about wind power, wave energy, tidal energy, or hydroelectric energy, we are talking about the secondary effects generated by the sun. When we consider hurricanes, tornadoes, and cyclones, we should realize the magnitude of the perturbation of the sun's energy; the storms have vested forces many, many times greater than the biggest H-bomb.

What we commonly mean by solar energy, however, is what we receive directly from the sun's presence. It is surely extraordinary how little we have done, except perhaps for sunbathing, to make effective use of solar energy in technological terms. The Greek inventor Archimedes (about 287–212 B.C.) is reputed to have set fire to the Roman fleet besieging Syracuse on the island of Sicily by the use of a burning mirror, but this laser beam of antiquity did not figure in subsequent armaments. In 1747 G. L. L. Buffon set up a battery of 140 flat mirrors in a Paris garden and ignited a stack of wood 200 feet away, to show that the claim for Archimedes could be sustained. A. L. Lavoisier, in this as in so many things, was an originator, inventing what we today call the solar furnace. Enclosing specimens of various substances in transparent quartz vessels, he placed them at the focal point of a lens 52 inches in diameter and used concentrated solar radiation to heat them in partial vacuum and controlled atmospheres. Lavoisier claimed (what is manifestly true) that the fire of ordinary furnaces seems less pure than that of the sun. This has been confirmed by its present-day use as the tool for high-temperature research. Just improvising with the parabolic reflectors taken from antiaircraft searchlights, research chemists have been able to concentrate solar radiation so effectively as to obtain temperatures as high as 6,332° F (3,500° C). The largest furnaces are those of Mont-Louis, France; Sendai, Japan; and Natick, Mass.

Producing salt by evaporation is an old and worldwide practice. Conversely, the removal of salt by distillation from brackish water or seawater gives potable drinking water. Now that solar energy has acquired respectability and is not regarded as the domain of cranks, there is also increasing use of its moderate-temperature application in the growing and drying of agricultural products, in cooking, and in heating buildings and domestic hot water supplies. Solar boilers can be used to operate ammonia-absorption refrigerators.

The difficulty about sunlight is its diffusion, its cutoff at night, and its irregularity through cloud interference. The total amount of solar energy reaching the Earth is about 7×10^{17} kilowatt-hours per year, more than 30,000 times as much as all forms of energy consumed in man-made devices. To collect, concentrate, convert, and contain this dispersed energy is difficult, if one considers the square miles of mirrors needed for energy of industrial proportions. The advent of satellites has encouraged ideas of using them, beyond the dispersing atmosphere, as relay stations.

But it is possible to convert solar radiation directly into electricity. The most efficient and convenient way is through the use of silicon photovoltaic cells, developed in 1954 by Bell Telephone Laboratories. A solar cell made of thin wafers of ultrapure silicon to which traces of arsenic and boron have been added can produce a direct current with a conversion efficiency as high as 16%. The effectiveness of solar batteries was demonstrated by their use in the U.S. satellite *Vanguard I*, which on March 17, 1958, was sent into orbit carrying 108 silicon cells. The satellite continued to transmit intelligible signals throughout the 1960s. Subsequent weather and communications satellites have had their outer surfaces covered by thousands of cells, some of which are always facing the sun and generating the power needed to charge their batteries and to operate their radio and television equipment. Larger space vehicles carry assemblies of tens of thousands of cells, mounted on panels that are folded into the rocket and then fan out when the desired course has been attained. To meet the larger requirements of space vehicles, thermodynamic power systems have been developed, using liquid metals as their working fluids, with metallic hydrides for high-temperature heat storage.

It is to be expected that one of the spin-offs of space exploration will be cheaper and more adaptable solar batteries for terrestrial use. With the variability of sunlight, a major problem is storage for use during the interruptions. The main region of usefulness for terrestrial applications of solar energy is between latitudes 30° N and 30° S where sunshine is plentiful but conventional energy sources are scarce and fossil fuel costs prohibitive. And it is in these areas that the less-developed countries are waiting for their "leap across the centuries" to a higher standard of living.

Like sunlight, winds are variable, and already there is articulate concern among environmentalists about a skyline disfigured by myriads of windmills. In moderation and in a congenial landscape windmills are quaint, but in mass arrays they are, to environmentalists at least, as unsightly as power pylons or industrial chimney stacks. There are areas where the winds are constant and where electricity could be generated if the cost benefit of storage and transmission could be accommodated.

Considerable attention has also been given to and success attained in harnessing tidal energy, but at high capital cost. Waves are another proposition. They are capable of generating enormous, renewable, and pollution-free energy supplies. The neglected science of waves is now being systematically developed, and technology is coming up with the kind of barrages that could be set up in the oceans. Hydroelectric power has laid emphasis on waterfalls and dams. There are still appropriate sites throughout the world, but they are not necessarily easily accessible to existing industries; again it is a question of storage and transmission. Useful advances have been made in small-scale units, like bulb generators, which can float in rivers and streams to exploit the currents.

Bioenergy can be exemplified at its crudest in the burning of cow dung, but there are vast quantities of natural waste that could be converted to convenient forms of energy, notably methane gas. Geothermal energy has manifested itself through aeons of time as geysers, mud pools, and vents of hot gases from the crust of the Earth. Wherever there is volcanic activity, there is geothermal potential, which can be exploited as it has been in Iceland, Italy, Japan, Mexico, New Zealand, the Soviet Union, and the United States. Heat from rocks can be derived in other areas as well.

The development of alternative sources of energy is imperative and urgent and fraught with contemporary social and economic convulsions. Even the shift from oil to coal, or the conversion of coal into oil, to meet the voracious appetite of the internal combustion engine, will mean an enormous relocation of capital resources. In the United States it will mean moving the coal industry out of the Appalachians, leaving behind a ravaged landscape and deprived people. In addition, the United Mine Workers, opposed to the mine owners, is poorly organized in the western United States, where development could shift. Billions of dollars will be needed in investments in synthetic oil and gas plants and new pipelines. The social geography of the United States could be radically changed.

The lead time in shifting from a fossil-fuel economy to alternative sources is considerable, and time is not on our side. The cost and abundance of energy affect not only transportation and the heating and cooling of our domestic environment. It affects not only the price but also the availability, the quality, and the variety of mankind's clothing and food and shelter, the basic necessities. It affects comfort, safety, and the purity of the environment. It affects political and economic stability within, and between, countries.

Energy is not a subject to be left to technicians, to vested interests, or even to politicians and scientists. It affects the common interests of all of us and requires our common understanding. And to acquire that understanding it is necessary to get away from the specialized textbooks of experts who treat the various forms of energy as, somehow, self-contained subjects. That is why this book deals with energy in all of its aspects. The book will have justified itself if it shows that energy is not OPEC and that the world need not be hostage to oil or to any particular form of energy. Human ingenuity can ring the changes.

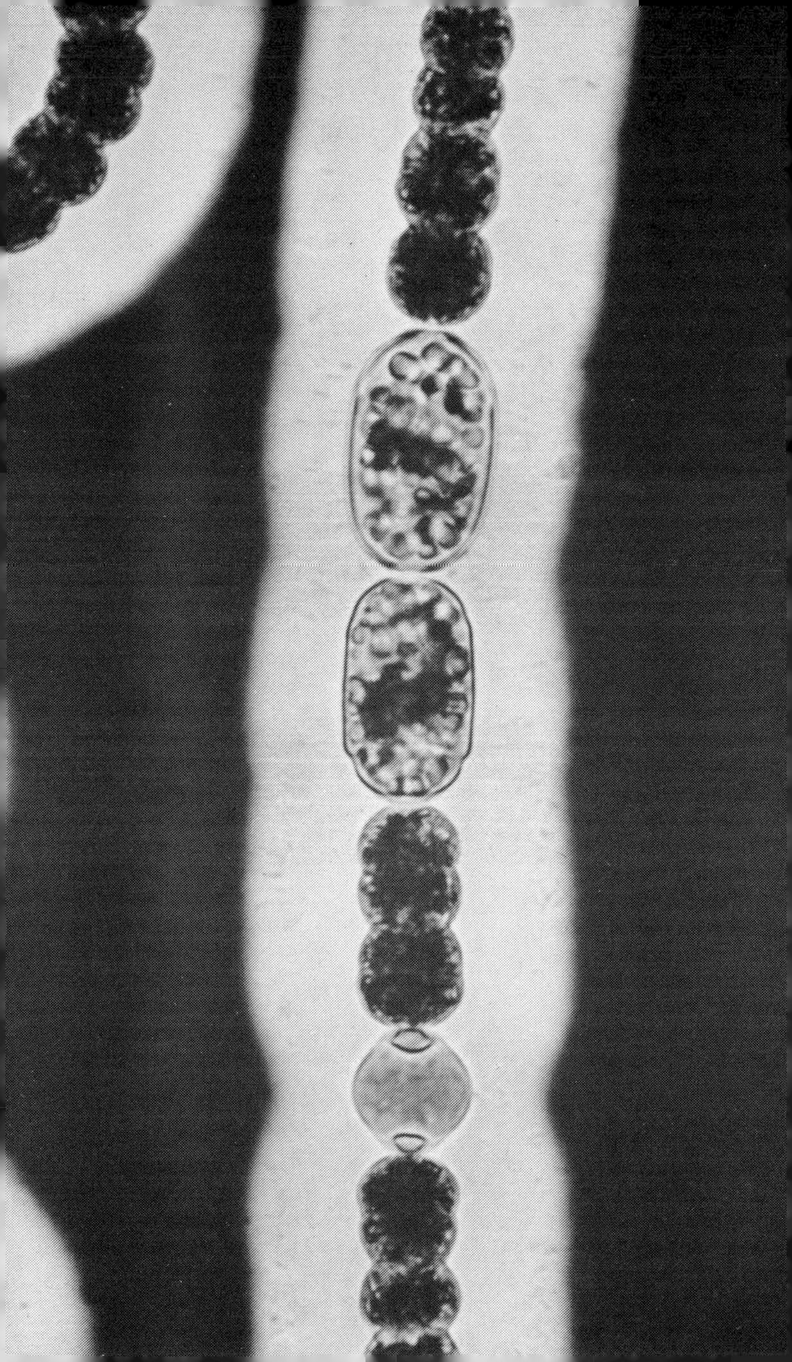

1.
The Energy of Life— Sun into Fossil Fuel

The creation of life and of usable energy called ATP; the relation between photosynthesis and the power of human muscles; respiration and nitrogen fixation in the biosphere; and the formation of coal and oil

All life depends on energy! To grasp this truth, it is necessary to go back to the beginning of what we call "life." For this exploration there are no means of direct observation. We have space machines but no time machines. Our space travelers, who have reached the lifeless moon, may go on to reach far-off planets, and we may further refine our electronic sensors and define more precisely the matter of Mars or the environment of Venus. But we already know enough to conclude that the conditions on other planets in our solar system will not give us a study of life making that compares with the Earth's history. It is a mathematical certainty to most scientists that in the immensities of the universe there are worlds with conditions identical to our own, but they are millions of light-years away. On Earth, even if we found locked away somewhere a pocket of primeval molecules, we could not be sure that this was a canned version of the "soup" that contained the original ingredients of life.

The Earth is about 4.5 billion years old. About 3 billion years ago its crust became relatively stable. There seems to be general agreement that the atmosphere was lacking in free oxygen, but that there were such gases as methane, water vapor, ammonia, hydrogen, carbon dioxide, carbon monoxide, and nitrogen—elemental groupings that could react with each other. In the laboratory these gases of the primeval atmosphere have been energized by electrical discharges and by ultraviolet rays, and the derivatives have been identified as amino acids and other organic molecules necessary for the process called "life." The vigorous reaction of these gases to ultraviolet light is important because, since there was no free oxygen in the original atmosphere, the ultraviolct radiations from the sun were not modified by ozone (O_3) and were therefore much stronger. Most of this chemistry probably took place in the upper atmosphere, and the molecules were precipitated by rain and collected in the seas, where they reacted upon one another. They formed aggregations that, in turn, became more complex. Some were more efficient than others in attracting elements out of the environment (like later creatures competing for food) and so grew at the expense of the less efficient. This was a primitive beginning for natural selection.

Meanwhile, changes were taking place in the atmosphere. Hydrogen, the lightest gas, escaped rapidly. Harold Urey, a Nobel Prize-winning U.S. chemist, estimated that hydrogen had declined to its present level 2 billion years ago. Lacking hydrogen, ammonia (NH_3) and methane (CH_4) were unstable and disappeared from the atmosphere. By that time, however, organic synthesis had taken place and the organic process had become self-sustaining.

The characteristic of a living system, according to John H. Northrop, a U.S. biochemist who shared the 1946 Nobel Prize for chemistry, is "to use energy to carry out synthesis of more of itself." This excludes from the definition of "living" any autocatalytic reaction (in which the chemical that facilitated the reaction emerges as a product of the reaction). It also rules out crystals, which grow by reproducing themselves but do so by giving away energy. On the energy gradient, a crystal runs downhill while the living organism climbs uphill.

ATP

The lack of oxygen was an important property of the early atmosphere. If oxygen had been present at those shotgun weddings of the elements, the precarious molecules would have ended simply in combustions. If life appeared without help from oxygen, it must have been supported by fermentation, which the French chemist Louis Pasteur described as "life without air." Fermentation makes energy available to a cell by breaking up organic molecules (those containing compounds of carbon) and releasing high-energy phosphates such as adenosine triphosphate (ATP). Certain forms of fermentation, such as those that produce alcohol, yield carbon dioxide as a by-product. The release of this gas into the atmosphere by anaerobic forms of life, which do not require oxygen, contributed to the evolution of later metabolic processes, including respiration.

After fermentation the next progression in metabolism was

the HMP (hexose monophosphate) cycle. It is essentially an anaerobic process that develops hydrogen from sugar with the aid of energy derived from ATP. It also releases carbon dioxide as a by-product. Half the hydrogen in the HMP cycle comes from water (H_2O). This cycle represented a relatively advanced stage (in millions of years) because it was "getting hydrogen the hard way" and indicates an era when practically all of the free hydrogen had fled the planet.

The third stage in this progression was probably photophosphorylation, the direct utilization of sunlight to produce ATP. This process requires the pigment chlorophyll (a magnesium porphyrin) to absorb light and the cytochromes (iron porphyrin proteins) to convert the absorbed external energy, sunlight, into the stored internal energy, ATP.

All living organisms derive their energy from sunlight, but only green plants can directly use sunlight to synthesize cellular components from simple sources such as carbon dioxide (CO_2), water (H_2O), and ammonia (NH_3). This process is called photosynthesis. Most other organisms must use the products of photosynthesis as food, either by eating plants or eating creatures that eat plants. The chemical reactions undergone by the components of food—mostly proteins, carbohydrates, and lipids (fats)—serve two purposes. They break down complicated substances into simple ones, with the release of energy that becomes available for the organism's activities, and they build up complex substances with the absorption or storage of energy. The breaking-down process is "catabolism," and the building up, "anabolism"; the two processes together are "metabolism." Living organisms can neither consume nor create energy; they can only transform it from one form to another. Nonuseful energy is returned to the environment as heat. Heat cannot perform energy functions in biological systems because all parts of the cells have essentially the same temperature and pressure.

One specific chemical compound, ATP, is used for energy exchange in all living organisms. It is the one form of energy adapted to cellular use. Think of a cell, any cell, as similar to an electric lamp. The energy to light the lamp can come from oil or coal, from the nucleus of an atom, or from a waterfall, but, whether thermal, nuclear, or kinetic, the energy has to be transformed into an electric current, the only form of energy that the lamp can use. Thus ATP is the body's grid supply of chemical energy. As it transfers its energy to other molecules, ATP loses its terminal phosphate group (P_i) and becomes adenosine diphosphate (ADP), or else it loses two of its phosphate groups (PP_i) and becomes adenosine monophosphate (AMP). These products can be reconverted to ATP by reacquiring phosphates.

Photosynthesis

Living green plants supply the sustenance of animals and human beings by fabricating carbohydrates and proteins in forms that creatures can digest and convert to the needs of their own metabolisms. The plant is a factory that derives its energy directly from the sun and its raw materials from the atmosphere, from water, and from chemicals in the soil or in the sea. The plant pigments have the capacity of trapping sunlight and transferring this imported energy to the chemical economy of the living process. About 2% of the sunlight falling on a plant is transformed into chemical energy. When light falls on the plant, the greater part of the energy is absorbed by small granules called "chloroplasts," which contain a variety of pigments including "chlorophylls." The chlorophylls transform the energy of light into chemical energy by a process that includes photolysis, the decomposition of water, and the activation of ATP. This, in turn, energizes the fixation of carbon dioxide (making it nonvolatile or solid) so that carbohydrate molecules are formed as sugar and starch. The splitting of water into hydrogen and oxygen releases a molecule of oxygen that becomes the atmospheric oxygen that we breathe. This free oxygen was not present in the original atmosphere, and thus our whole respiratory apparatus depends entirely on what the photosynthetic action on the land and in the sea has produced.

In the next stage of photosynthesis the carbon dioxide plus four atoms of hydrogen become carbohydrate plus water. That is the beginning of a flow sheet that is more complex and more efficient than any industrial chemical installation and that, with the introduction of nitrogen, creates the amino acids that in turn become the proteins.

In a living unit as simple as a bacterial cell, there are five thousand different kinds of protein, all starting from sugar and ammonia. Each has a special function. Some are knitted together to form membranes that enclose the cell and form its internal partitions. Others are long chains, curled up like skeins of wool. These are globular proteins, and they make up most of the jellylike component of the cell. These proteins within the cell are energetic. Some act as enzymes (biochemi-

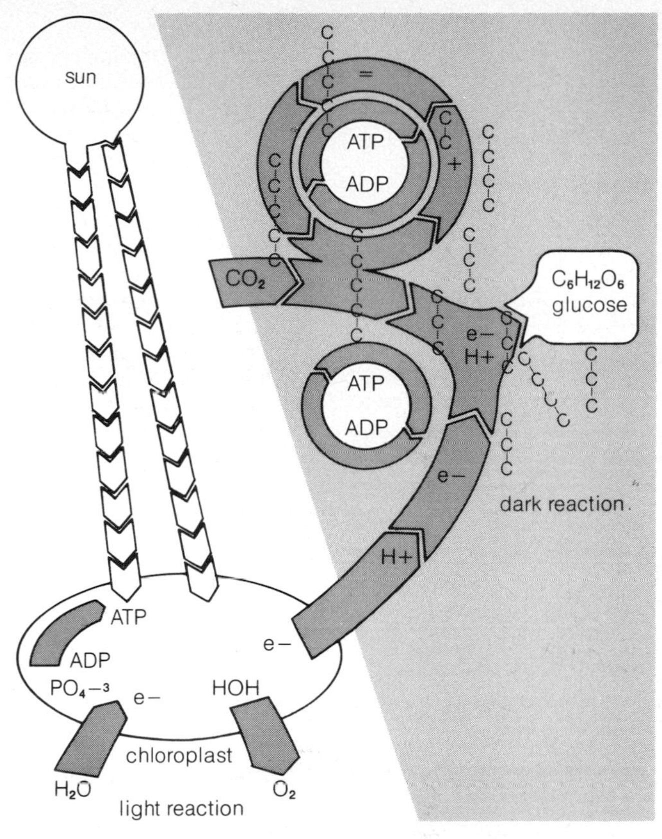

During photosynthesis light from the sun splits water molecules, freeing oxygen; by means of chemical reactions within the plants, the remaining hydrogen plus carbon dioxide from the air are incorporated into glucose—a simple sugar or carbohydrate. At the same time, light energy causes ADP to react with a phosphate group and become ATP—the energy-rich substance that powers glucose synthesis. The carbon chains represent various intermediate stages in these cyclical light and dark reactions.

cal catalysts), which decompose nutrient molecules that are ingested into the cell and rearrange their elements into amino acids and nucleotides (organic compounds consisting of sugars, purines or pyrimidines, and phosphate groups).

A bacterial cell will ultimately reproduce itself by dividing into two cells if it is immersed in a solution of sugar, phosphate, and ammonia. The sugar and ammonia molecules are relatively simple, and yet the energetic system within the cell can rearrange the atoms, contrive from them twenty kinds of amino acids and four nucleotides, and assemble them in the correct order to become the proteins and nucleotides necessary for reproduction. The amino acids are the units that link together into the polypeptide chains, which, in turn, cross-link to form proteins. The nucleotides form the energy-carrying ATP and the nucleic acids deoxyribonucleic acid (DNA) and ribonucleic acid (RNA), which store and transfer the genetic code. This code determines the structure of cells within the living organism as well as the hereditary traits passed on from generation to generation. This activity can go on as long as a carbohydrate source is available, and it can only be provided by plants with their photosynthetic capabilities of using sunlight.

Muscle Power

For tens of thousands of years the only working form of energy *Homo sapiens* and his precursors could use was muscle power. Today hundreds of millions of people in the developing countries of the world still have to rely on their own, or their animals', muscles. Their fuel is the food they eat, and, in terms of cost efficiency, they are using the most expensive form of energy. A peasant, or his ox, on minimum rations is (pricing the kilocalories) using energy at a unit cost twenty times that of electricity from a nuclear-powered station. Or, to express muscle power in a mind-boggling example, if a modern transatlantic liner were a galley ship, it would take 3.5 million slaves pulling on oars to row it from New York City to Cherbourg, France, in five days.

Muscle has sometimes been compared to a heat engine. It is true that both systems produce work by oxidizing combustible materials (combining them with oxgyen), but the parallel ends there. Muscle works at constant temperature and resembles an electric motor driven by a battery. The chemical energy of the battery is converted into mechanical work without prior conversion into heat. The consumption of zinc in the battery is analogous to the consumption of foodstuffs in the muscle.

Muscular energy is derived from the combination of carbohydrates with atmospheric oxygen to form carbon dioxide and water. Carbohydrate is stored in the muscle as glycogen, which the body can form from protein and fat in the diet as well as from other carbohydrates; and oxygen is brought to the muscle by the circulating blood. This oxidation is a complex process involving many different enzyme systems. It can take place only slowly, and the energy liberated is not in a form that can be used directly by the contractile proteins, those that form the pulling tissues of the muscles. The energy derived from these oxygen-using (or aerobic) processes is therefore used to drive forward a complex system of anaerobic chemical reactions that lead eventually to the synthesis of creatine phosphate (CP) and of ATP. The latter is readily converted in the presence of the appropriate enzyme into ADP plus energy, and the energy is handed on to the contractile proteins. The store of ATP is quickly regenerated by driving the above reaction backward, the energy required being derived from the hydrolysis (chemical decomposition involving the addition of elements of water) of creatine phosphate. ATP and CP are present in quite appreciable quantities, and they constitute an important local store of energy. This permits muscles to work at a high rate for a few seconds, or until the local store is used up.

All muscles can outstrip their oxidative processes during severe exercise, and the local blood supply may not be able to transport enough oxygen for the needs. Under these conditions energy can be obtained from glycogen in the body not only by oxidizing it but also by splitting it anaerobically into lactic acid. Surplus lactic acid is carried by the bloodstream to the liver and to inactive muscles, where about 20% of it is oxidized to form carbon dioxide and water. The energy obtained is used to rebuild the remaining 80% into glycogen. In this fashion the metabolic load is spread over the whole body instead of being confined to the active muscles.

When a muscle becomes active, it needs extra oxygen. The blood vessels in the muscle, therefore, open up so that more of the total blood flow is diverted into the active region. If the exercise is at all severe, there will have to be compensating changes in other parts of the cardiovascular system. Blood vessels in the skin and viscera close down, and the heart beats faster and more forcefully in order to circulate the blood more rapidly. The active muscles produce extra carbon dioxide, which stimulates sensitive chemoreceptor cells in the brain. This causes the breathing to become deeper and quicker so that carbon dioxide is more effectively washed out of the blood and more oxygen is absorbed.

These compensating mechanisms make it possible to transport a much larger quantity of oxygen than usual, though, of course, they have their limits. Even the most athletic man breathing air cannot absorb more than about 4.5 quarts (4.5 liters) of oxygen per minute (this is about twenty times the resting value), and in order to distribute this oxygen to the tissues each half of the heart must pump about thirty quarts of blood per minute. Since each quart of oxygen consumed yields about 5,000 calories of energy, the maximum human rate of continuous energy production is about 1.5 kilowatts. Assuming an efficiency of 25%, this corresponds to production of mechanical work at the rate of 0.5 horsepower. This is about equal to the value obtained by direct measurement in a man during strenuous rowing. This limit is set by considerations of oxygen transport rather than by intrinsic limitations in the muscles. If a high concentration of oxygen is breathed instead of air, the oxygen is more effectively absorbed into the bloodstream, and by this means it is possible to work at higher rates without discomfort. The beneficial effect of breathing oxygen only appears when the physiological mechanisms are under strain, as during hard exercise or at high altitude. Under normal conditions there is no detectable benefit.

The Biosphere

The biosphere is that part of the Earth in which life exists. It is the living envelope, while the atmosphere is the gaseous envelope, and the lithosphere is the rocky crust. Its existence depends on that fraction of solar energy that is converted by photosynthesis.

The biosphere probably had its beginnings 2 billion years ago with the evolution of marine organisms that could fix solar energy in organic compounds and also split the water molecule to release free oxygen. This oxygen accumulated for hundreds of millions of years, slowly creating an atmosphere that screened out the most destructive rays or modified them. New species of plants and creatures evolved that derived more energy by respiration, which speeded up the processes. Evolution adapted the fauna and flora into a living-together system that not only conserved energy and the mineral nutrients necessary for the life processes but also recycled them, releasing more oxygen and making possible the fixation of more energy and the support of still more life. The 350 million cubic miles (1.5 billion cubic kilometers) of water on the

Earth are split by photosynthesis and reconstituted by respiration once every two million years, and the oxygen released by the plants into the atmosphere is recycled every two thousand, so that the air you inhale now may contain oxygen atoms that the Roman general Julius Caesar exhaled as he crossed the Rubicon.

There is more to plant life than the creation of organic compounds by photosynthesis, however. Plant growth involves a series of chemical processes and transformations that require energy. The part of the process that releases carbon dioxide is called respiration.

Respiration

Respiration is a round-the-clock process. In a forest setting marked changes of carbon dioxide concentrations can be measured. The average distribution of carbon dioxide in the atmosphere is about 320 parts per million. When the sun rises and photosynthesis is switched on, there is a rapid decrease of carbon dioxide as it is used by the trees to make organic compounds. On the midday shift the rate of respiration rises, and the consumption of carbon dioxide declines to 10 to 15 parts per million below the daily average as measured at treetop level. At sunset photosynthesis is switched off while respiration continues, so that the carbon dioxide concentration close to the ground may exceed 400 parts per million.

A fair estimate is that each year the land areas fix 20 to 30 billion tons of carbon into organic compounds. The amount of carbon, in the form of carbon dioxide, consumed annually by phytoplankton in the oceans is about 40 billion tons. Both the carbon dioxide consumed and the free oxygen released by the photosynthesis of the phytoplankton are largely in the form of gases dissolved near the ocean surface. Within the ocean system the released oxygen is consumed by sea creatures, and their ultimate decomposition releases carbon dioxide back into solution. At any given moment the amount of carbon dioxide dissolved in the surface layers of the sea is in close equilibrium with the concentration of carbon dioxide in the atmosphere as a whole.

Nitrogen Fixation. Nitrogen forms 79% of our air. It is an essential element in the formation of amino acids and of proteins and is also the basis of nitrocellulose, nitroglycerine, and trinitrotoluene (TNT), all explosively energetic. As an atmospheric gas, however, nitrogen is inert. It has to be fixed into some chemical compound before it becomes available to plants or animals. In nature nitrogen can be made to react with oxygen or the hydrogen of water by cosmic radiation, by meteor trails, or by lightning, which momentarily provides the high energy needed for the flash welding of the molecules. Some marine organisms and some microorganisms are exceptional in being able to fix free nitrogen, and many legumes have nodules on their roots containing microorganisms of the genus *Rhizobium* that provide the green plants with growth-stimulating nitrogenous compounds.

At the end of the nineteenth century the British chemist Sir William Crookes painted a doomsday picture of a world in which food production would collapse because of the lack of fixed nitrogen. This was the time when the Chilean guano reserves were the main source of both fertilizers and explosives. The guano deposits, excretions of the vast seabird population of that part of the world, were being heavily exploited, and Crookes foresaw their exhaustion in the efforts to feed multiplying populations of the industrial countries.

During the early 1900s two German physical chemists, Fritz Haber and Karl Bosch, invented the catalytic fixation process. In their original procedure atmospheric nitrogen and hydrogen were passed over a catalyst (nickel) at a temperature of about 900° F (500° C) and a pressure of several hundred atmospheres to yield ammonia. In most modern manufacturing plants the hydrogen for the basic reaction is obtained from methane, a natural gas. Methane and steam react to produce a gas rich in hydrogen. Then atmospheric nitrogen is introduced. The accompanying oxygen is converted into carbon monoxide by partial combustion with methane. The carbon monoxide then reacts with steam. Carbon dioxide is removed to be used in a side process to convert ammonia to urea, which is $CO(NH_2)_2$. What is left of the carbon monoxide is converted into methane. Then nitrogen and hydrogen combine at high temperature and pressure, in the presence of a metal oxide catalyst, to form ammonia. Ammonia reacts with oxygen to become nitric acid. Nitric acid can be combined with ammonia to produce ammonium nitrate, a widely used fertilizer. Meanwhile, back at the root nodule, a little bug is somehow doing all that for the legume. Nitrogen-fixing bacteria can, with the help of an enzyme (nitrogenase), accomplish at ordinary temperatures and pressures what in chemical engineering requires hundreds of degrees of temperature and thousands of pounds of pressure.

As is obvious, the fixation of nitrogen requires an invest-

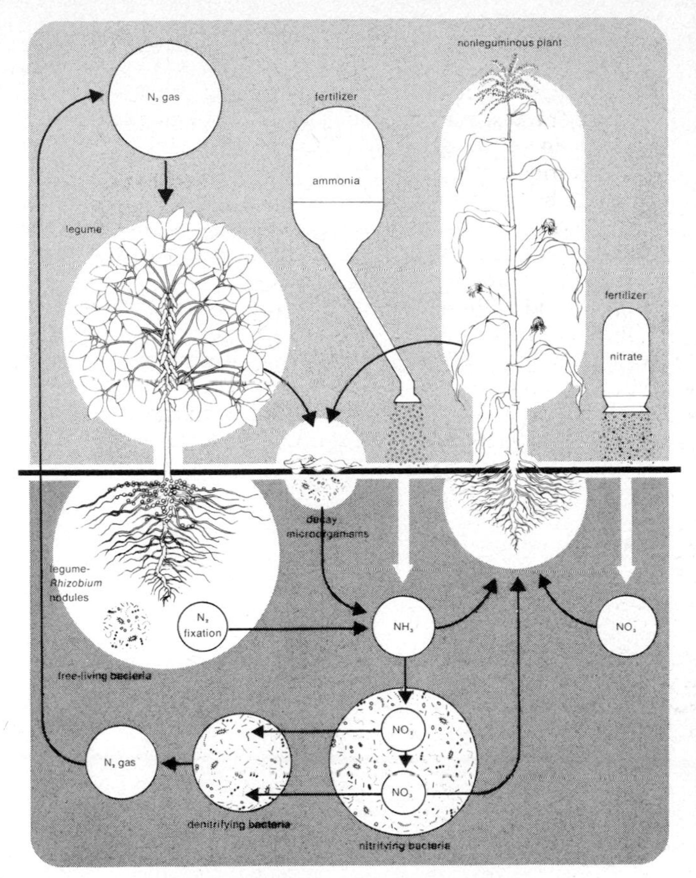

The nitrogen cycle includes bacterial conversion of atmospheric nitrogen and organic compounds to ammonia. Man contributes to soil nitrogen in the form of fertilizer ammonia and nitrates. Ammonia is removed from the soil by nitrifying bacteria and living plants. Nitrates from bacteria or fertilizer may also be used by plants or returned to gaseous nitrogen through bacterial action.

ment of energy. First, each molecule of atmospheric nitrogen has to be split into two atoms of free nitrogen. This step requires at least 160 kilocalories for each 24 grams of nitrogen. When two nitrogen atoms combine with six atoms of hydrogen to form two molecules of ammonia (NH_3), 13 kilocalories are released. Thus the two steps need a net input of 147 kilocalories. In industrial terms that means about 6,000 kilocalories per kilogram of nitrogen fixed. Crop trials indicate that the yield increase produced by a kilogram of nitrogen fertilizer amounts to about the same number of food calories. According to U.S. biochemist C. C. Delwiche, since 1950 the amount of nitrogen fixed annually for the production of fertilizer has increased approximately fivefold until it now equals the amount fixed by all terrestrial ecosystems before the advent of modern agriculture. By the year A.D. 2000 the industrial fixation of nitrogen may well exceed 100 million tons annually. That is four times the amount of fixed nitrogen delivered to the Earth by rainfall.

Roots of the common bean possess bulbous nodules that result from infection with the symbiotic bacterium Rhizobium phaseoli. *Swelling occurs as root cells become filled with these nitrogen-fixing organisms.*

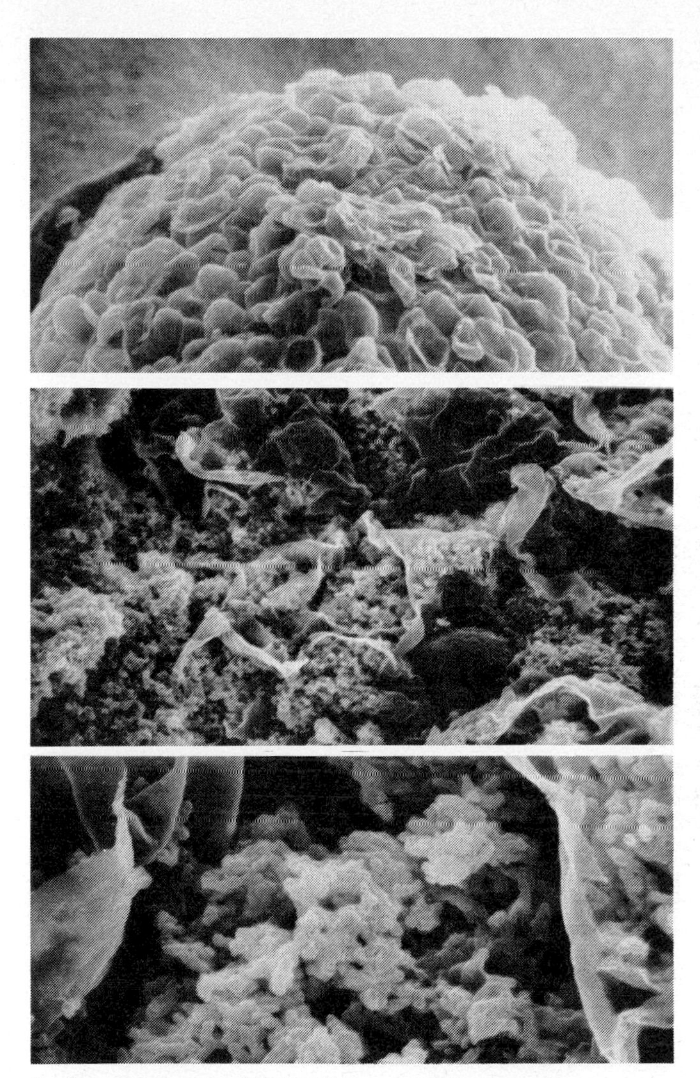

Soybean root nodules filled with nitrogen-fixing bacteria are seen in a sequence of electron micrographs showing the surface of a single soybean root nodule (top), a nodule cut open to reveal the interior (center), and a dense mass of Rhizobium *spilling from an opened root cell (bottom).*

A nonindustrial mass producer of fixed nitrogen is the legumes, which include peas, soybeans, and alfalfa. Such plants have gone into partnership with bacteria that lodge in the nodules on the plant roots. Symbiotic nitrogen fixing of this kind has a special need for trace elements such as molybdenum or cobalt. (They are rather like the presence of a metal oxide as the catalyst in the industrial process.) Molybdenum is directly incorporated in the enzyme nitrogenase. As little as two ounces of molybdenum per acre will suffice.

Apart from the symbiotic lodgers there are free-living bacteria, such as *Azotobacter*, which supply fixed nitrogen to grasslands. In an aquatic environment an active nitrogen fixer is the blue-green algae.

Denitrification. Denitrifying bacteria are important parts of the nitrogen cycle. It was their identification at the end of the nineteenth century that panicked Crookes and others. These bacteria, in the absence of oxygen, are able to use nitrate or nitrite ions to accept electrons from organic compounds and thereby oxidize them. In those reactions, the energy yield is nearly as large as it would be if pure oxygen were the oxidizing agent. For example, when using oxygen the conversion of glucose yields 686 calories per gram of glucose. In microorganisms the reaction of glucose with a nitrate ion yields 545 calories if the nitrate is reduced to nitrous oxide and 570 calories if it is reduced all the way to nitrogen gas. It is this function of restoring nitrogen to the atmosphere that caused concern in the nineteenth century: if it went on unheeded, soils would become bereft of nitrogen, and crops would be crippled.

Indeed, denitrification does continue on a big scale, but it is no longer regarded as scary. On the contrary, the worry is that denitrification may not be coping with the man-made, or man-induced, nitrogen excesses produced by artificial fertilizers and by the vast acreages of legume crops. If it is not somehow regulated, the balance of the biosphere will be upset by a buildup of nitrates, nitrites, and ammonia. The problem of nitrogen disposal is aggravated by the nitrogen contained in the organic wastes from human and animal excretions—once part of the discreet ecology of farmyard manure and domestic "night soil" but now totally distorted by the sewage disposal of our vast urban centers.

It is possible to have too much of a good thing. Unless fertilizers and nitrogenous wastes are carefully managed, rivers and lakes could become loaded with runoff nitrogen, and in the waterways and in the groundwater system the nitrogen concentration could make the water unfit, indeed dangerous, for human consumption.

Photosynthesis is the means of capturing and storing energy from the sun. It is the only way that energy can be made available for the processes we call "living"—growth, repair, reproduction, and all the other dynamic processes, including the energizing of the brain cells to function as thought or imagination. Chlorophyll is the "stuff that dreams are made of." Perhaps our brains, drawing on the chemical energy and electrical energy provided for us by green plants, will devise ways of synthesizing food from the elements but—as can be gathered from the relatively simple example of fabricating just one essential, usable nitrogen—the cost would be astronomical. In photosynthesis the energy is free.

Not only are the carbohydrates created by photosynthesis, but also the amino acids, proteins, lipids (or fats), and other organic components of plant cells are produced by this method. In the process, chemical bonds are broken between oxygen, carbon, and hydrogen, and new bonds are formed in products that include oxygen and organic compounds containing the elements. This excess energy, provided by light, is stored as chemical energy (the give and take of electrons) within the products. That stored energy, provided by photosynthesis, is what is being talked about when either the world's "food problem" or its "oil problem" is discussed today.

The Making of Coal and Oil

Having gone back all the way to the laboratory of creation, perhaps it will now be possible to get what has been called "the energy crisis" into perspective. It all began those billions of years ago with the processes that converted the energy of the sun into ATP. The chlorophylls and other pigments of plants used the energy of the sun to convert carbon dioxide, water, and minerals into oxygen and energy-rich organic compounds as food for creatures great and small, including *Homo sapiens*. This process also gave rise to organic mineral deposits—the hydrocarbons of coal, oil, and natural gas.

Coal was formed from the remains of living trees, shrubs, and other plants that flourished during periods of relatively mild and humid conditions. Although some coals were deposited 400 million years ago during the Silurian period, most were formed about 250 million years ago in the Lower and Upper Carboniferous periods. Conditions then favored the growth of huge tropical seed ferns and giant nonflowering trees in vast swamp areas. As the plants died they fell into the swamps, which excluded oxygen and encouraged anaerobic decay. The vegetation was changed into a slimy material called peat. Some peat was brown and spongy, and some was black and compact, depending on the degree of rotting. The sea advanced over such deposits, and mineral sediments were laid on top of them. Under pressure the peat dried and hardened to become low-grade coal (lignite). Further pressure and time created bituminous coal, in which a twenty-foot thickness of the original plant material was compressed into one foot of coal. Even more extreme pressures, resulting from the folding of the Earth's crust into great mountain ranges, produced the hardest and highest-grade coal, anthracite. The quality of coal is rated by the amount of fixed carbon in relation to the amounts of moisture and volatile matter (matter turned into gases as a result of heat).

Considering the vast amount of time and effort that man has put into finding and extracting oil and natural gas, he is still surprisingly ignorant about their origins. There are certain accepted facts: oil and gas consist of compounds of biological origin; most types of petroleum contain porphyrins, complex hydrocarbon compounds derived either from chlorophyll or from hemin (the red coloring matter of blood); the lipid fractions (fats and waxes) of organisms provided a large part of the source material for oil and gas; and in present-day conditions it is possible to find petroleumlike hydrocarbons in recent marine sediments. Furthermore, petroleum is commonly associated with sedimentary rocks that have been deposited under marine conditions.

The organic material from which petroleum has been derived was probably single-celled planktonic plants, such as diatoms and blue-green algae, and single-celled planktonic animals, such as foraminifera. Those primitive forms of life were abundant more than half a billion years ago and could have formed the source of the petroleum found in the Precambrian and Lower Paleozoic rocks. They could also have provided the materials found in younger rocks as well. After the death of the cells, the preservation of organic materials requires immediate burial in fine-grained, clay-sized sediments. Such matter must be protected from oxygen, which would break up the molecules and completely destroy it. In appropriate sediments diagenesis proceeds. Diagenesis is the

Predictions vary as to when the Earth's finite oil supply will be depleted, but experts agree that there is an urgent need to develop alternative energy resources.

sum of all the ways in which chemical changes turn loose sediments into rocks. It is a relatively low-temperature, low-pressure system. This is consistent with the fact that petroleum can be found in commercial quantities at depths from 30 yards (about 30 meters) to more than 7,500 yards, indicating that pressure is not an indispensable requirement. Nor is high temperature—higher, that is, than about 200° F (100° C).

Crude oil is a mixture of thousands of different chemicals, which range from extremely light gases to semisolid materials such as asphalt or paraffin wax. Approximately 8% of the oil is found in rocks of the Palaeozoic Age (from 225 million to 570 million years ago), 63% in rocks of the Mesozoic Age (65 million to 225 million years ago), and 29% in those of the Tertiary and Pleistocene ages (from 10 thousand to 65 million years ago).

The Babylonian word *naptu* (which became the Greek word *naphtha* for an "inflammable liquid") was used as early as 2000 B.C. and meant "to flare up." In the Babylonian temple tablets it was applied to omens and ascribed to angry gods who fired the liquid with lightning. The sound of natural gas issuing from rock fissures was described by King Tukulti-Ninurta of Babylon (885 B.C.) as "the voice of the gods speaking out of the rocks."

It is, or ought to be, a sobering thought that in man's insatiable demands for fossil fuels he is rifling the geological vaults in which the fuels were laid down hundreds of millions of years ago. They were discovered quite late in man's tenancy of this planet. Coal was burned in China as long ago as 100 B.C., but it was not mined in Europe until the early thirteenth century A.D. and did not reach flamboyant proportions until the Industrial Revolution in the eighteenth and nineteenth centuries. The invention of the steam engine and the steam locomotive created a sharply increased demand for coal, which was needed as fuel to heat the steam boilers and also to carbonize iron. Before then, wood had been used to produce the charcoal that in the smelting process transferred to iron ore the carbon required to produce pig iron.

The real beginning of the oil industry was the drilling of Drake's Well at Titusville, Pa., in 1859 and the development of the nearby Rockefeller refineries. These developments made possible the internal combustion engine and the automobile. During the past century industry has vented out of chimney stacks and car exhausts some 360 billion tons of fossil carbon into the atmosphere. Throughout the world 2.9 billion tons of coal and coal equivalent are wasted annually. If present trends continue, the cumulative requirements between now and the year 2000 will amount to about 110 billion tons of coal, 200 billion tons of oil, and 50 trillion cubic meters of natural gas.

2. The Big Bang— From Anaxagoras to the Twentieth Century

Theories of the origin of the universe; the physics and cosmology of Anaxagoras; space and time in Newtonian and Einsteinian relativity; "black bodies," the quantum theory, and the principles of uncertainty and complementarity

The Big Bang is the reigning theory of the origin of the universe; of all the galaxies; of all the stars, including the sun; of all the planets, including the Earth; and of all life, including the brain that thought up the theory of the Big Bang. Even the most lusty supporters of alternative theories, like the Steady State or the Continuous Creation, have had to acknowledge the weight of the evidence.

The principal ingredient of the Big Bang was not matter but radiant energy—from radio waves, through light waves, to gamma rays. Indeed, these original rays are still around. It was the discovery of this residual background radiation, pervading the entire universe with a temperature of 3° K, that reinforced the claims for the Big Bang. The discovery was made by U.S. physicists Arno A. Penzias and Robert W. Wilson in 1965 and checked out by Robert Dicke and his colleagues at Princeton University. This had been theoretically anticipated by George Gamow, Ralph A. Alpher, and Robert C. Herman, whose Big Bang model assumed that the early universe was not only fantastically dense but fantastically hot. They argued that low-temperature radiation would persist as present evidence if only it could be detected and measured. It was.

It is estimated that the Big Bang occurred 15 billion years ago. In a countdown of a thousandth of a second, it triggered off a process that released the energy and generated the heat that were to create all the forces and all the matter to fill all the immensities of space with radiations, elements, gases, and materials. Isaac Newton surmised, more than 280 years ago, that matter would "convene, some of it into one mass and some into another so as to make an infinite number of great masses, scattered great distances from each other throughout all that infinite space." At Point Zero, somewhere in space, all those billions of years ago, the Big Bang went off. At one second after blast-off the temperature was 5,000,000,000° K. At this time the universe was flooded with a light that was denser than matter. As the cosmic fireball cooled off and the universe expanded, the variety of forces and particles that we are learning about today began to emerge.

The universe went on expanding and matter went on "convening." The configurations of the nebulae, stippled by stars, were not static patterns on a celestial tapestry but were moving; they were moving outward, farther and farther from Point Zero and from each other. It was like (but with forces and dimensions billions of times greater) the effects of a nuclear bomb: the explosion of the core, the fireball, the radiation, and the eruption and spreading of the mushroom cloud. It is all there on our contemporary newsreels in miniature. A film, however, can be reversed—like showing a diver leaping from a springboard and then showing him traveling upward onto the springboard again.

What modern scientists are trying to do is a backward rerun. They are tracking back to events that they can identify billions of light-years away. This is a long way, since light travels at a speed of 186,000 miles (300,000 kilometers) per second and in a year would traverse about 5.9 trillion miles. When a signal is picked up from a source, therefore, say 900 light-years away, it means that, here and now, we are recording an event that happened 900 years ago—like getting a news flash that William the Conqueror has landed in England.

In the runback scientists get to Point Zero and, in compulsive logic, to "singularity" when existence and time ceased to exist. That is frustrating. Scientists can hope to measure forces and particles, but they cannot measure nothingness. Even short of "singularity," they face a quandary embodied in Albert Einstein's equation, $E = mc^2$, that emerged in the development of his special theory of relativity. The equation, which means that energy equals mass multiplied by the square of 186,000 miles per second, was cataclysmically demonstrated by the release of energy by matter in the nuclear bomb, and the high-voltage accelerators have shown how energy creates matter. From what primordial atom came the pent-up energy that filled the immensities of space with radiations, galaxies, stars, and planets?

Back to Anaxagoras

Anaxagoras shared the quandary. He lived in Greece between 500 and 428 B.C. and died just before Plato was born. Socrates, born forty years after Anaxagoras, turned to the latter's book to learn the laws that governed the universe.

Anaxagoras asked the questions that are still being asked 2,500 years later: What is the nature of matter? What are the fundamental properties of the substances that make up the world around us? Is matter discrete or is it continuous? What laws govern the changes that matter undergoes? What keeps objects together and endows them with their properties? Anaxagoras combined strict and detailed observation with the method of logical analysis being formulated in the Greece of his time.

From his direct observations Anaxagoras could see the transmutation of matter and at the same time its indestructibility. The growing tree is nourished by matter from the soil and rain from the clouds. When it is cut down it is burned as a log. Part of the log rises as smoke into the clouds, from which rain descends into bodies of water in which salts and earth-precipitates form. Part of the log remains as ashes that crumble into soil and nourish plants that, in turn, become food for animals. To Anaxagoras it seemed that in these transformations the matter of the log eventually would pass through every state of every substance in the material world and, indeed, the universe. This idea of the conservation of matter led Anaxagoras to accept infinite variety and continuous change in the structure of matter as a universal process, and he tried to explain it in stable categories and to find the theories to justify it.

Anaxagoras observed only limited processes of decomposition and concluded that an infinite set of ultimate substances existed, his elementary particles being what today might be called molecules. Long after his time, when it was learned how to break the bonds that held individual molecules together, atoms appeared to be the elementary particles. Atoms in turn were stripped down to yield their nuclei and electrons. With higher expenditures of energy the nuclei can be broken down into the subnuclear particles of today.

Anaxagoras chose "life" as his criterion of elementarity. He produced a theory of the unity of matter that did not require two separate assumptions, living and nonliving. According to him, one of the inherent and eternal properties of all matter, animate or inanimate, was the organic nature of its basic constituents, his "uniform substances." This notion of living elements was most convenient because it enabled Anaxagoras to construct his molecular concept of the cosmos and to adopt "reason" as the source of order in the universe, for which Aristotle commended him.

Because Anaxagoras's concepts have a relevance to contemporary debates on cosmology, energy, and the structure of matter, they are worth examining. He believed that he had to account not only for general change (and, at the same time, the eternal conservation of matter) but also for the emergence of an infinite variety of objects from any given object (as from a burning log). He reasoned that the only way an infinite variety of stable entities can exist in a finite object is for the entities to be infinitesimal in size, because an infinite number of entities can be contained in a finite volume if they are infinitesimally small. He was, therefore, using the term *infinitesimal* (2,500 years ago) in precisely the sense that it has been used since the seventeenth century: magnitudes whose measure is larger than zero but smaller than any arbitrary small number.

Anaxagoras also was the first on record to realize that it is not enough to determine the nature of material components that make up the universe but that it is also necessary to provide an explanation for the motion and change that matter undergoes. He did not simply say, as other classical Greek philosophers did, that matter behaves the way it does because it is its nature to behave that way. He sought to find a unifying principle that would account for the behavior of each and every part of the universe for all eternity. He was saying that the goal of dynamics is the discovery of one "law" that governs all phenomena of physical reality. He was trying to find a principle that would fulfill the following conditions: (1) It must be one, not many; (2) It must account for all motion and changes of state that take place and must suffice to explain and give direction to all physical change; (3) It must be thoroughly logical, rational, and devoid of any mythological, capricious, or theological character; and (4) It must account for the marvelous order that seems to exist in the universe and for the harmonious functioning of all parts of the physical world. Postulating that such an entity really exists, Anaxagoras gave it the name *nous*—"reason," or "mind."

To him this name was not just poetic nor anthropomorphic; it was the logical extension of his understanding of living matter. As a present-day parallel scientists observe the cellular structure and the convolutions of the human brain and infer that it is constructed of atoms, but at the same time they distinguish the anatomy and physiology from the thought processes, which, somehow, those atoms generate—the mind. Just as the mind of each living being pervades his whole body, Anaxagoras's cosmic mind pervaded all nature. This was consistent with his theory of the composition of matter from living elements. *Nous* in his system admitted of no disorder and controlled all things. The universe thus was an infinitely great, structured organism, not simply a mechanism involving forces that we can now recognize but which Anaxagoras did not conceive.

To Anaxagoras the matter of the universe lay muddled in chaos for eons. Then *nous* took over. The first step separated heavier from lighter matter by churning. He had observed that eddying liquids or gases draw to their center objects denser than the medium. Ships are sucked into a whirlpool at sea; dust is pulled into the vortex of a windstorm; and houses and trees are wrenched into the eye of a cyclone. *Nous*, he said, turned the handle of the cosmic churn and out of the universal random mixture brought most of the dense matter of the universe to the center, leaving the air and gases around it. This was only a rough separation that produced a gross structure on which refinements proceeded after the churning had stopped. On the Earth geological formations were created, vegetation arose, animals and humans came into being; to Anaxagoras civilized living was an extension of *nous*'s manipulation of molecules. The paradox was that *nous* should have been the perfection of the process, but Anaxagoras had to assume that it was the beginning. This is akin to the quandary of present-day cosmogonists.

The Fourth Dimension

Time is the "when" of common experience, and there is also the "where," that is, position in space. In the abstract the latter is difficult to grasp. To the real estate agent "space" is what he pegs out as a "lot" and sells. To those concerned with space exploration, "space" extends outside the confines of this planet. We cannot ordinarily "imagine" space; we tend to relate the concept to material objects as they appear in our sense experience. We are led to discuss space in terms of markers, of the relative position of objects. In those terms, space is conceived in a physical context, constrained by the observation that material bodies occupy different positions. This is a convenient notion when all the positions of bodies are described in relation to one body, such as the Earth. This is what continues to make Euclidean geometry so satisfactory for so many practical purposes. The point, the straight line, and the plane often are accepted as having a self-evident character.

This apparently logical self-sufficiency, however, has become inconvenient for physicists. In ordinary mechanics every event seems determined by "place" (position relationship) and "when" (temporal relationship). To two-dimensional and three-dimensional space (geometry) was added the inescapable consideration of time, but space still could be accepted as a separate and perfectly serviceable dimension in classical physics. This reconciliation was possible because of what is now recognized as the illusion of "simultaneity," that is, that when we receive the news of an event (say, the motion of a planet) instantaneously by the agency of light, we see it happening here and now.

The faith in absolute simultaneity was destroyed by the laws revealed by electrodynamics. James Clerk Maxwell, a nineteenth-century Scottish physicist, found that disturbances in an electromagnetic field travel with a definite velocity, the speed of light, whatever their wavelength. Light is only a special case of all those disturbances that are wave phenomena. Heinrich Hertz, a German physicist, later discovered that ordinary electric disturbances—"sparks"—could produce an electric field some distance away. This led to the detection and transmission of radio waves with their enormous variety of wavelengths. These vary from the long waves used in telegraphy to the short waves used in television and radar. In addition to wavelength a basic property of electromagnetic radiation is frequency, which is the number of oscillations per second. Frequencies are measured in hertz (cycles per second).

The visual wavelengths, which we call "light," are those that excite the atoms of the materials that form the retina of the eye. The longest normally give us the experience of red; the intermediate ones give us yellow, green, and blue; and the shortest ones give us violet. Longer than visible waves but shorter than radio waves are the infrared (heat-producing) waves. Shorter than the visible violet is ultraviolet (which gives us suntan). Shorter still are X rays, and even shorter are gamma rays from the nucleus of the atom.

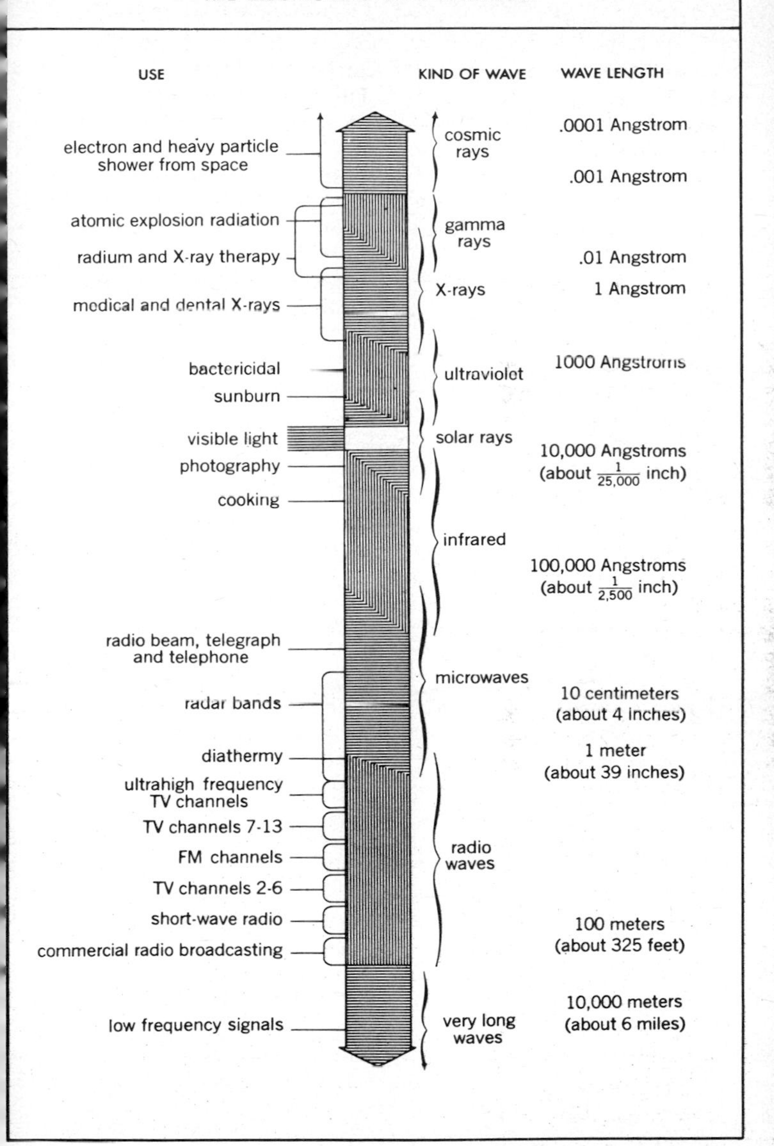
THE ELECTROMAGNETIC SPECTRUM
USE
KIND OF WAVE
WAVE LENGTH
electron and heavy particle shower from space
atomic explosion radiation
radium and X-ray therapy
medical and dental X-rays
bactericidal
sunburn
visible light
photography
cooking
radio beam, telegraph and telephone
radar bands
diathermy
ultrahigh frequency TV channels
TV channels 7-13
FM channels
TV channels 2-6
short-wave radio
commercial radio broadcasting
low frequency signals
cosmic rays
gamma rays
X-rays
ultraviolet
solar rays
infrared
microwaves
radio waves
very long waves
.0001 Angstrom
.001 Angstrom
.01 Angstrom
1 Angstrom
1000 Angstroms
10,000 Angstroms (about $\frac{1}{25,000}$ inch)
100,000 Angstroms (about $\frac{1}{2,500}$ inch)
10 centimeters (about 4 inches)
1 meter (about 39 inches)
100 meters (about 325 feet)
10,000 meters (about 6 miles)

What is remarkable is that this vast range of waves—of different frequencies, differently excited, and differently received—all behave in accordance with the laws that Maxwell propounded. One property is that some waves can be bent by reflection; for example, radio waves are mirrored by the ionosphere so that they can overcome the curvature of the Earth. Waves of higher frequencies, such as those needed for television, do not act so conveniently. They are limited by horizon range or have to be purposefully redirected by communications satellites. They all travel, however, at the velocity of light.

An important application of radio waves is radar, the means by which they are used to locate distant objects. A short pulse of radiation is bounced off a target, thereby giving information about its direction and distance. The direction is simple, just that in which the signal is sent out. Distance is measured by inference from the interval of time between the instant of the transmission of the pulse and the instant of the reception of the echo. Since we know that radio waves travel with the speed of light, the interval multiplied by the speed of light and divided by two (half the round trip) gives the location of the target; the continuing signals, bouncing back, give the speed of the quarry.

Radar has provided a method of measuring distance in which one does not use a yardstick. No standard meter or yard is employed. What one does is to measure an interval of time and then multiply this by a constant quantity, the velocity of light. Thus time can be used to express distance. This, of course, is familiar in astronomy; to avoid awkwardly large figures in discussing the distance to the stars in miles, such distances are given in light-years, based on the distance (about 5.9 trillion miles) that light travels in a year. One also can speak of "light-microseconds," based on the distance light travels in a millionth of a second (about 300 yards). A "light-millimicrosecond" would be a thousandth of that unit.

In ordinary, mundane affairs, as in clocking a race or setting a radar speed trap, the speed of light is not likely to matter very much (one would scarcely fight a ticket for speeding by disputing the measured distance in terms of five-millionths of a second). When considering the nature of the universe, however, such aspects of time and distance become important. Once it is understood that time is distance and vice versa we recognize that the star we see is not a here-and-now phenomenon but that its light has been traveling to reach us for billions of years; thus it is billions of billions of miles away, and what we have been witnessing, in our present time, is an event that took place in the remote past. When space and time become thus inseparable—when one cannot think of one without the other—time ceases to be one-dimensional as in classical mechanics, and space-time becomes the fourth dimension.

Classical Relativity

If different observers are to make observations on the same set of phenomena, their findings cannot be compared accurately unless each observer's results are convertible into the terms used by other observers. It is like translation from one language into another, which can convey an exact meaning provided the respective foreigners agree on their dictionary definitions—that *vache*, in French, means "cow" and not "horse," or, more aptly, that 13 *heures* means "1 P.M.," or that *kilometer* means "five-eighths of a mile." Mathematicians call this a transformation from one system to another and express it by means of a set of "transformation equations."

Classical (pre-1900) mechanics could get along satisfactorily with such transformation equations in what can be called "Newtonian relativity." A railway track, for example, runs across the Great Plains—no curves and no hills to interfere with constant velocity. But saboteurs have mined the track and explode two bombs at different times, each at a different place. The explosions are observed by two different people, the stationmaster at a depot and a traveler on the train. They have identical watches, and as the train flashes through the station, proceeding at constant speed, each notices that it is precisely 12 o'clock. Each observer is also using instruments to measure distance. When the bombs go off, the stationmaster will put down two numbers X_1 and X_2 as giving the distances of the two explosions from the depot and T_1 and T_2 as the times of the explosions. The man on the train speeding toward the explosions will put down X'_1 and X'_2 for his distances and T'_1 and T'_2 for his times.

In classical mechanics, as we have seen, time and distance were two separate absolutes, unchanging and the same for all observers: If the stationmaster finds that the distance between the two explosions is ten miles, the man on the train must also so find it, and if the man on the train says that the time interval is 35 minutes, the stationary observer must agree. Furthermore, in this frame of reference, time intervals and distance intervals between events must be the same for all observers no matter how they are moving relative to one another. This leads to a simple transformation equation by which a description of any event in a moving system can be derived from that in a fixed system. First, all the T values (time as noted by the stationary observer) are agreed to be equal to the T′values (time as noted by the moving observer). Second, to adjust the X′ value of any event, that is, its distance from the moving observer, it is only necessary to multiply the velocity of the traveler by the time of the occurrence of the event and to subtract this from the X value of the event, that is, the distance of the event from the stationmaster. The mathematician expresses this transformation in the following terms: $X' = X - vT$ and $T' = T$, where v is the speed of the man on the train. All that has been stated in the transformation is that the fixed clock and the moving clock keep identical time and that the distance of the points on the railroad track is less for the traveler than the distance of the same points from the stationmaster by the distance the man on the train has moved from the station. That, it would seem, makes common sense.

Science, however, imposes another demand, the principle of invariance, which insists that transformation from one system to another must not trifle with a law of nature just because it is convenient to do so. Such mathematical transformation certainly takes no liberties with Newton's three laws of motion and the principles of the conservation of energy, and until the middle of the nineteenth century nobody felt called upon to question it. Then Maxwell, following the experimental work on electromagnetism by the British physicist Michael Faraday, produced his equations showing that all radiation (including light) is propagated in the form of waves. What he showed was that the speed of light is constant, independent of the motion of the source and of the motion of the observer. This means that if a lamp is flashed at the railway station and both the stationmaster and the passenger measure the speed of light as it moves in the direction of the train, they should both get the same result. (The same would apply if the beam originated from a star.)

If one applies the transformation equation used in the "sabotage" illustration, it would follow that the traveler would find the speed of light somewhat less than that calculated by the stationary observer. This is consistent with ordinary experience: if a car is traveling at 40 miles an hour and a

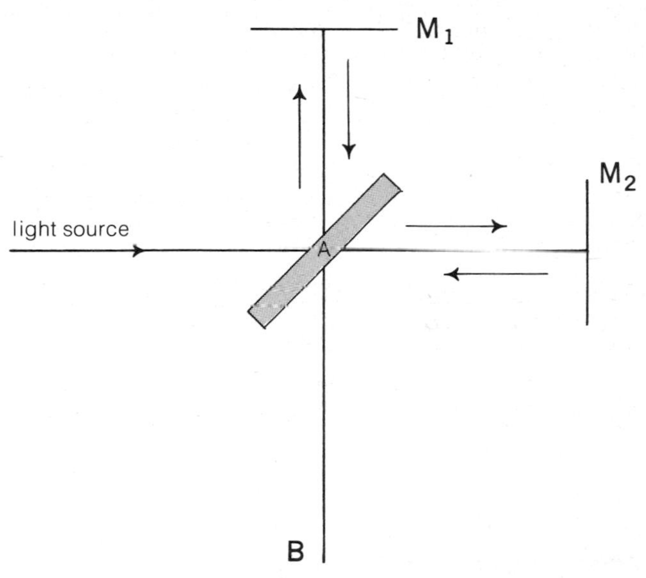

It is probable that no instrument has more profoundly influenced modern physics than the Michelson interferometer; the results of one of its first applications, the Michelson-Morley experiment, gave evidence that light travels at constant speed in all inertial systems of reference—the experimental foundation for Einstein's special theory of relativity. The device consists of an incompletely silvered mirror, A, that divides a beam of light waves—transmitting half to a fixed mirror, M_2, and reflecting half to a movable mirror, M_1. Reflected back, the waves recombine, revealing any interference with one another in a visual pattern of light and dark bands called fringes. An observer looking along BA sees M_1, the virtual image (M'_2) of M_2, and the fringe configuration. By counting and measuring the fringes when M_2 is moved away from A, precise wavelength determinations, study of spectrum lines, determination of refractive indices of transparent materials, and measurement of very small distances and thicknesses are possible.

second car is traveling at 50, the latter will draw away at 10 miles an hour. This seems manifestly true of moving bodies and, as the nineteenth-century physicists maintained, anything that Maxwell said to the contrary must be wrong.

Then, in 1887, an experiment was devised by the U.S. physicists Albert Michelson and Edward Morley to test whether there was a difference in the speed of light in different directions. If the speed of light was different for a moving observer, it should have been possible to show it by measuring the speed of light and comparing the result with the value given by Maxwell's theory to determine the speed of the Earth through space. In other words, it should be possible to detect absolute motion.

If the speed of light is considered in terms of a beam moving in the same direction as the Earth's motion and of another beam moving at right angles to that direction it should have been possible to detect absolute motion, and if the absolute concepts of space and time were right, then, certainly, Maxwell must be wrong. Michelson and Morley experimentally measured the two beams and showed a result that meant that either the Earth was standing still or that Maxwell must be right.

That contradiction of classical physics only confused the situation until 1905 when Einstein proposed a solution. He started from two fundamental assumptions. The first is that all laws of nature must remain unchanged when one transforms from a given system to another system that is moving with uniform velocity with respect to the first system; this is the principle of invariance. If the transformation equations are based on the correct concept of space and time and the so-called law does not conform to this principle, then it must be amended so that it does. The second assumption (which Einstein accepted as a fact, in view of the Michelson-Morley experiments) was that the speed of light is constant to all observers. From these assumptions he went on to derive a new set of transformation equations that unify space and time—the space-time continuum.

Einstein's Relativity

Although it is like trying to measure one's waistline in millionths of an inch, we can consider classical space and time absolutes and relativity's space-time by looking again at the "sabotage" illustration. In classical theory the distance between the two explosions should be the same for both the moving and the fixed observers, and the same should hold for the time interval between the incidents. In the theory of relativity this is no longer the case because distances and time intervals vary with the speed of the observers. There is, however, a space-time interval between events that remains constant for all observers. No matter where they are or how fast they are moving they are all "within the law."

As a consequence of this theory, simultaneity and distance are not accepted as invariants. If two occurrences are observed (call them A and B), one observer may insist that they happened simultaneously, another may maintain that A happened before B, and a third may say that B happened before A—and each will be faithful to his own frame of reference. Similarly, an observer may measure a definite distance between events, but another may say that they coincided in space.

Of course, all this squaring, multiplying, and subtracting is not going to settle a ball game argument among the referee, the linesman, and the spectator; nor has it relevance to everyday experience because the adjustment of the transformation equation becomes imperative only when the speeds involved get close to the velocity of light. Such speeds are now obtained in the giant accelerators that impel particles of the atomic nucleus. At those speeds time and space get hopelessly mixed up, and the only measure that still remains the same for all observers is a certain space-time interval.

Einstein's equation $E = mc^2$ expresses the equivalence of energy and mass. It says that energy equals mass multiplied by the square of the velocity of light. It says that a moving body, whether it is a particle or the sum of many particles, gains mass (amount of matter) proportionate to its increase in speed. Conversely, as has been shown with cataclysmic violence in the explosion of nuclear bombs, energy is released proportionate to the loss of mass. In the case of the atomic or fission bomb the sum of the mass of the elements produced by splitting is less than the mass of the original atom. The complete fission of a pound of uranium would produce about 2.5 million times more energy than is produced by the combustion of a pound of coal. In the case of the thermonuclear bomb, the fusion of lighter atoms to form heavier atoms—hydrogen into helium—releases even greater energy than the fission of heavy elements. The process of converting four hydrogen atoms into one helium atom, which is understood to be the principal source of the energy of the sun, involves the conversion of 0.7% of the mass into energy. For the purposes of a bomb the reaction of a deuteron ("double-hydrogen") and a triton ("triple-hydrogen") to form a helium nucleus plus a neutron is accompanied by the conversion into energy of 0.4% of the combined mass of the deuteron plus triton.

At the beginning of the twentieth century, therefore, there was an upheaval—the classical physics of Newton had been found wanting. Newton's three laws of motion could not completely meet the observed facts. The law that says that the force acting on a body is obtained by multiplying the mass of the body by its acceleration, for example, can have no ultimate meaning if the mass of the body changes as it moves; there is no way of knowing which value of mass to use in expressing Newton's law. His law of gravity involves the masses of bodies and the distances between them, and in Einstein's theory both these quantities vary from observer to observer. Which mass? And what distance? Einstein, after producing his special theory of relativity in 1905, set out to formulate his general theory, to reconcile relativity and gravitation.

Since the gravitational problem is concerned with the mass of the body, Einstein directed himself to this and to the fact that mass appears in two different identities in classical mechanics. Classically, mass is the quantity obtained when the force exerted on a body is divided by the acceleration which that force imparts to the body. This quantity is referred to as the "inertial mass" of a body. But "mass" also classically refers to the quantity called the "gravitational mass" of a body in the formula for the force of gravity between the two bodies. Experimentally, the inertial mass and the gravitational mass are numerically equal, although the two quantities are physically unrelated. Einstein was convinced that this was no terminological coincidence; that there was a "principle of equivalence."

Let us imagine Newton in the seventeenth century sitting under an apple tree that is growing out of solid English soil. He watches an apple fall to the ground and perceives a profound truth: a force from the Earth is pulling down the apple and all other bodies with the same acceleration regardless of their mass. He calls this "the force of gravity." Let us now imagine Einstein, in the twentieth century, in an elevator suspended in space. It is a curious elevator because there is an apple tree growing in it and its upward acceleration is 32.2

Principle of Equivalence

feet per second each second. An apple falls to the floor, which is not the massive planet attracting the apple but a platform rushing upward to meet it. Newton's gravitational pull downward has become Einstein's acceleration rate upward, and there is no way by which either observer can experimentally prove whether he is at rest in a gravitational field or moving in an accelerated system.

A straight line, we were once told, is the shortest distance between two points; but if we draw a line on a sphere, such as a model of the Earth, in order to link two points, the shortest distance along the surface will be the arc of a great circle. (This "great circle" is now familiar as the shortest route from our airport of departure to our airport of arrival.) Carl Friedrich Gauss, a German mathematician, had provided mathematical tests for determining curvature or flatness. If the curvature is zero, the surface is flat and the shortest distance is the shortest distance of classical geometry. If it is positive (greater than zero), however, or negative (less than zero) the surface is either elliptical or hyperbolic, and the shortest distance is a different kind of "straight" line. This work of Gauss was further developed by Georg Friedrich Riemann, a German geometer, who applied it to curved spaces of any number of dimensions.

This non-Euclidean geometry was available to Einstein when he needed to apply it to the concept of four-dimensional space-time. He assumed that the space-time construct of the universe is not flat but curved. He mathematically formulated the theory that a freely moving body (one that is not subjected to the push or pull of any other physical object) moves in a Euclidean straight line unless a gravitational field is present. If there is a gravitational field, the body moves in a curved orbit that is the four-dimensional, or space-time, shortest distance between any two points. By this method the gravitation field acting on a body is no longer considered as a force but as a curvature in space-time. Einstein worked out the invariants that would apply under all transformations from one system to another. Astronomical observations of the planets and of the sun at the time of its eclipse have supported Einstein's "curvature of space."

The Black-Body Revolt

Another set of principles was emerging at the beginning of the twentieth century. It began with what may be called "the black-body revolt" against Maxwell's electromagnetic wave theory. That theory had itself been a scientific revolution. According to the wave theory, light is a periodic electromagnetic vibration propagated through (etherless) space. In the case of visible radiation the wavelength is expressed in angstrom units (1 angstrom equals one ten-millionth of a millimeter). Red colors have wavelengths of about 7,000 angstroms, and deep violet colors have wavelengths of about 3,500 angstroms, but as Maxwell showed, radiation is not restricted to the visible range. The electromagnetic spectrum extends from radio waves a mile long (from crest to crest) to gamma rays with wavelengths equal to a fraction of an angstrom.

A second basic property of radiation is frequency. There is nothing complicated in this idea; if one watches a buoy bobbing up and down on the sea waves, the number of bobs in a given time gives the frequency of the waves. In the case of visible light the frequency (or vibrations per second) is very large.

It was this question of frequencies that showed the weakness of the wave theory as expressed by Maxwell; it did not fit all the observed data. If a piece of metal is heated, it turns a dull red, then bright red, and then white. As it gets even hotter, it emits violet rays and ultraviolet rays. There is, however, a special case known as "black-body radiation." This can be understood if one considers what happens if radiation falls on an opaque surface. Part of the radiation is absorbed, and part is reflected. If the surface were perfectly white, it would all be reflected. On the other hand, if all radiation is to be absorbed, a perfectly black surface is necessary. A black body, however, not only absorbs heat energy but also emits it—not by reflection but by accumulating heat so that the black body gets hotter and hotter and radiates it over the entire spectrum.

To reproduce, experimentally, the black-body effect, it is possible to enclose an intense heat source in a container (furnace) with a tiny hole. The experimenter can measure (by a spectroscope) the components of the radiation being emitted through the hole. As the temperature of the source increases so does the amount of energy radiated and the intensity of violet frequencies. Maxwell's wave theory could deal with the first finding. It takes care of the observation that, if the absolute temperature is doubled, the total amount of energy emitted per second from the pinhole will be 16 times greater; and if the temperature is tripled, the rate of emission will increase by a factor of 81, and so on. It also agrees with the experimental finding that the wavelength of the maximum intensity varies inversely with the absolute temperature (the hotter, the shorter).

Beyond that the wave theory could not cope with observed facts. Pushed to the limit the theory would have produced ultraviolet catastrophe. (Scientists, who usually deplore sensationalism, can nevertheless indulge in sensational terms. All that this portentous description means is that a black body should emit all of its energy in the ultrashort-wavelength region in one violent outburst.)

Every attempt to reconcile the experimental observations with the theory failed, and in 1900 the German physicist Max Planck showed mathematically that as long as it was assumed that a black body emits radiation continuously in the form of waves the ultraviolet catastrophe was inescapable. He suggested that radiation is not emitted continuously but in little discrete packets (quanta), each "photon" or quantum of light having its own wavelength and frequency. This was not a resurrection of the corpuscular theory of light that the nineteenth-century Newtonians had tried, and failed, to reconcile with the wave theory. Planck's "photon" could have no mass when at rest. It could have existence only when it was moving with the speed of light; it would vanish, or become part of an ordinary particle, when it came to rest. This posited substance, which was conceived to be as unsubstantial as a shadow, did not satisfy Einstein. He insisted that not only was radiation emitted discretely in the form of photons but also that it continued as photons at all times. He drew his support from, and at the same time explained, the photoelectric effect.

Einstein's experimental support came from the work of the German physicist Heinrich Hertz on radio waves. Hertz used an induction coil to produce sparks between two metal knobs and found that he was able to get better sparks if he irradiated the knobs with ultraviolet light. The air between the knobs became electrically charged when the metallic knobs emitted electrons under the influence of ultraviolet light.

According to the intensity of the rays, or the sensitivity of the materials used, the emission of electrons can be produced by visible light as well. This is the explanation of the photoelectric cell, which changes light into electric currents—the flow of electrons. Consider what this has meant in the entertainment industry (among thousands of other applications). The photoelectric cell permits the use of the sound track of movie films. The television camera (for black and white or color) consists of a mosaic of photosensitive cells, which, responding to the gradations or color frequencies of light, converts them into electric signals. These can be transmitted, to be picked up by a television receiver and converted into a beam of electrons in the television tube. This beam impinges on the flat internal surface of the tube, which is coated with chemicals. Those, in their turn, convert the cathode rays back into light frequencies that viewers can see. Or the frequencies produced by the original photons (the parcels of light that hit the camera mosaic) can be stored as untransmitted signals on a videotape, just as voice vibrations, converted into electron patterns by a microphone, can be stored magnetically on recording tape. When the magnetized patterns are reactivated by a magnetic pickup head, the original electron process is reproduced. A photon, therefore, is "something" that punches electrons out of the atoms of a material and frees them to move.

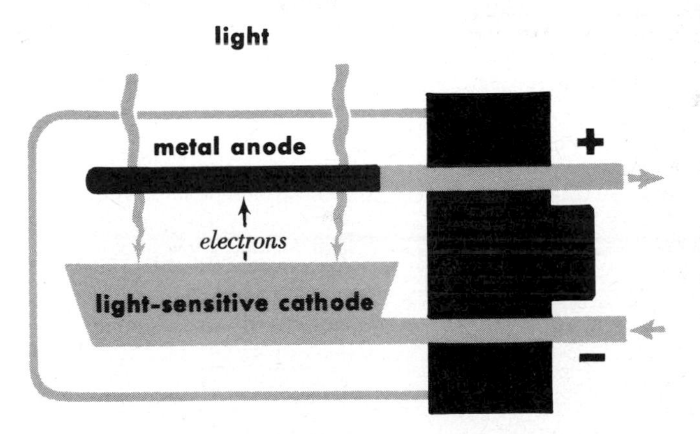

The photoelectric cell is the device that puts electricity to work. One type, the photoemissive cell, is used to open doors automatically. A beam of light shines on the photoelectric cell, containing a light-sensitive cathode and a metal anode that are enclosed in a vacuum or gas-filled tube. When light strikes the cathode, electrons jump from it to the anode, breaking the beam of light, causing the cell to activate a circuit that starts a motor, which opens the door.

The speed of the emitted electrons depends only on the frequency of the light used. The higher the frequency (or bluer the light), the faster the electrons move. If the intensity of the light is reduced the speed of the electrons is not affected; only fewer electrons are punched out. Einstein was able to produce a formula that can be extended to cover the behavior of light at all times and accounts for the measured energy of the emitted electrons.

The relationship between energy and color can be expressed in this way: to find the energy of a photon one multiplies the frequency of the photon's motion by the number 6.625 divided by one and twenty-seven zeros—1,000,000,000,000,000,000,000,000,000. (Why? Take Planck's word for it.) This is written as 6.625×10^{-27} or, in mathematical shorthand, *h*.

Planck's constant, *h*, is one of the indispensables of modern science. If it were successfully challenged (and, since nothing is scientifically sacred, attempts have been made to do so) it would upset most of the present theories of the universe and undermine the quantum theory.

Quantum Theory

To most people the quantum theory is as intimidating as Einstein's theory of relativity, of which it was once said: "Only three people understand it—Einstein, Bertrand Russell, and God." The quantum theory is now the working tool of thousands of scientists. Since it is so fundamental to physics, from the nature of the heavens to the nature of the atom, it is advisable to get on nodding terms with it.

Quantum mechanics has the reputation of being much more abstract than the classical mechanics that reigned supreme before it. The reason for this is that the objects of study of quantum mechanics appear to be much less directly related to everyday life than those of classical mechanics.

The word to emphasize here is *appear* because the "everydayness" is deceptive. For instance, Newtonian mechanics considers first "mass points" and then "rigid bodies" that are supposed to be composed of "mass points." One could take a billiard ball as exemplifying a rigid body, or, with some abstraction, a mass point. It is usually accepted, without too much questioning, that Newton's mass point has fewer properties than a billiard ball that one could actually hold in the hand. Unlike the billiard ball, it has no color nor can it be thought of as being warm or cold. Moreover, it does not, like the billiard ball, occupy a space because (as Euclidean geometry insisted) a point has no extension. This used to worry classical philosophers who occupied themselves a great deal with the problems posed by such notions as the extension (or lack of it) of a point. Such arguments can be conveniently ignored in Newtonian mechanics; one writes down equations for the extensionless, colorless, heatless mass point and interprets what one has put down in terms of a mental picture of the billiard ball or something like it.

If one wants to be more realistic about a billiard ball, one treats it as a Newtonian rigid body—with a lot of reservations. For one thing, the ball is not absolutely rigid. Even if it is solid ivory (or if it were made of steel, like a ball bearing) it can be compressed and distorted when submitted to enormous forces. If one looks at it under a microscope, one sees that it is not bounded by a smooth geometric surface. Atomic theory states that it consists of particles ("mass points"? or "rigid bodies"?) with spaces in between each, and all in quite violent motion. Nevertheless, what Newton says about rigid bodies (gravity, momentum, acceleration, and so on) is so well reflected by the observed behavior of billiard balls that we quite easily forget about the abstraction of Newtonian mechanics and equate its propositions directly to the things we observe in everyday life. "It makes sense," we say, meaning that it is consistent with our senses.

If the laws of quantum mechanics are abstract, so are the laws of Newtonian physics. This relation between Newtonian mechanics and the illusion of everyday objects is exactly paralleled by the relation between quantum mechanics and the atomic system. Those who "live with the atom" and have, by continual experimentation, become familiar with the "behavior" (the systematic results of measurements) of atomic systems, may learn to accept the statements of quantum mechanics as providing a completely coherent and rational model embodying the main features of what has been consistently observed. What is missing, however, is the accompanying mental picture in terms of familiar objects.

In hindsight it is not as surprising as it seemed at the time of major atomic discoveries that the entities we classify as atoms, elementary particles, photons, and so forth should behave not at all like billiard balls. It is the preconception that somehow they ought to do this that produces the main stumbling block for many who seek to grasp quantum mechanics. Basically, the "model" provided by quantum mechanics is mathematical, and it can only be appreciated fully in mathematical language. Again, this is not really surprising because all the observations on atomic and subatomic systems are essentially quantitative in nature, and if they are to be related one to another the tool for doing so has to be mathematical. It is helpful, however, even for the specialist, to have some sort of mental picture. And that brings us back to "waves."

In classical physics, waves are a subject of study distinct from mechanics. Clearly, some waves, like those on an ocean or in a vibrating crystal, can be interpreted as the combined motions of a large number of Newtonian particles. But in looking at an ocean-wave motion one ignores a great deal of possible information about the constituent particles in the wave.

It is no accident and not at all inappropriate that in popular language the same word, *wave*, is used, for instance, to describe the spread of an epidemic. We can think and talk of a "wave of influenza" without putting names and faces to the individuals involved or without tracking down each individual virus particle or without following the movements of a particular human carrier who spreads it. In other words, we have an epidemiological abstraction that we can accept without demanding identification of all of its items. In the same way, waves are a subject that mathematicians can independently study and that can be accepted (in some parts of physics) for the description of phenomena that are not reducible to the combined motion of many particles. A radio wave, for example, indisputably travels through intervening space since a receiver anywhere along the path intercepts it. But there is no need in the space for any particles whatsoever to make the transmission possible.

As has been pointed out, the abstractions forming the subject of Newton's light theories and Maxwell's theories are entirely different. What the founders of quantum mechanics found was that the behavior of atomic and subatomic structures required a mathematical scheme that could accommodate both the particle concept as used in Newton's mechanics and the wave concept as used in Maxwell's theory. The entity called an electron cannot be visualized by picturing a billiard ball; what one needs is part of Newton's "mass point" picture and part of the "wave" picture. Since anything that is visible to us in everyday life cannot be conceived as any mixture of a particle and a wave, such entities as electrons do not possess any close similarities with objects in our common experience.

Sir Joseph John Thomson, a British physicist who discovered the existence of electrons, without, of course, seeing them, said that he did not care whether they were pushed around by red-nosed pixies; what he was interested in was their behavior.

Pixilated Particles

Let us see, therefore, how pixilated physicists themselves are in their indirect observation of particles or light waves or whatever invisible matter they are studying. Suppose a metal is heated in a vacuum. Under suitable conditions it may be observed that a negative electric charge leaves the surface of the metal and moves in the vacuum (in a radio or television tube). This charge can be collected on other metal surfaces or be made to come into contact with all kinds of different pieces of apparatus. There are many ways of obtaining experimental evidence that both the emission of this charge from a heated surface and its acceptance by any piece of apparatus put in the vacuum container for that purpose do not proceed in a continuous, smooth way. Empirically, the charge appears to be carried by a large number of separate parcels (consistent with Thomson's notion of electrons), each the same size.

The U.S. Nobel Prize-winning physicist Robert Millikan showed that minute oil drops behave as if they carry just one, two, three, or more of these elementary units of charge, and he magnetically measured the unit. Thomson further noted that a beam of such units could be deflected in a magnetic field in such a way as would be predicted on grounds of Newtonian mechanics and the laws of magnetism if each charge were associated with a fixed unit of mass. The quantum physicist, without committing himself to a description of the "parcel," accepts "something" which he says has "charge e and a mass m_e." (This is like finding the fingerprints but declining to draw a picture of their missing owner—in this case, the electron.) The physicists in the early days of the electron's discovery were not so coy; they pictured the electron as a minute particle.

Nobel Prize-winner Thomson's son, G. P. Thomson, also a Nobelist, was partly responsible for upsetting the picture of a pixie-pushed particle. In a series of experiments streams of electrons were made to pass through an array of regularly spaced holes, and their direction of flight after emergence from this lattice, or sieve, was inferred. (*Inferred* is an over-punctilious scientific word. Only indirect evidence in the form of spots on a photographic plate was seen, but because some agent apparently produced the spots the posited electron's path could be "inferred.")

It was found that the electrons were emerging not in one beam and not in all directions at random. What did emerge were separate, regularly spaced beams. What was seen was something well known in optics as an "interference pattern." Such a result was not predictable by classical mechanics. On the other hand, classical optics prominently features the notion that two or more waves can, according to circumstances, either reinforce each other or cancel each other out. A single beam of light, for example, in passing through a grating or lattice splits into a set of regularly spaced beams just like the beam of electrons.

In optics the existence of an interference pattern is understood to require that the several parts of a wave that take

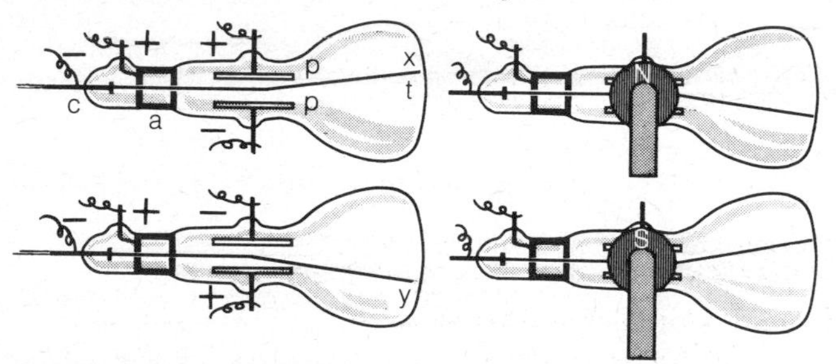

J. J. Thomson's experiments in electric and magnetic deflection suggested the existence of a particle much smaller than an atom. He built an apparatus in which a positive electrical charge on a two-plate anode (a) drew rays from the negatively charged cathode (c), which passed through the anode and struck the far end of the tube (t) enclosing a moderate vacuum. When positive and negative charges were put on the plates (p) in the tube, the rays deflected toward the positive charge, and the fluorescent spot shifted from center to either x or y. The rays were also deflected by the poles (N, S) of a magnet placed around the tube. In his experiments, Thomson produced a magnetic deflection, then applied electric charges just strong enough to cancel the deflection. From all the forces used, he calculated the ratio of the charge on one particle to its mass, which was more than a thousand times greater than it was for hydrogen ions in electrolysis, indicating that the particle's mass was correspondingly lower than that of a hydrogen ion or atom.

different paths through a lattice form part of the same general wave, matched in frequency and phase. The pattern could never be produced by allowing pulses of light to pass at random, each through a different gap in the grating. This, in terms of G. P. Thomson's experiment, apparently rules out any possibility that one electron went through one hole and another through another hole and came together again to produce the interference pattern. The inescapable conclusion had to be that each electron passed, in part, through all the gaps in the grating and, so to speak, interfered with itself. This indicated that any notion of localizing the electron had to be discarded. At the time, these experimental results were held to be paradoxical—for the very reason that localization was already part of the habit-formed mental picture.

But the apparent contradictions did not end there. Let us go back to the spots on the photographic plate. The picture of an electron emerging from the lattice as a spread-out series

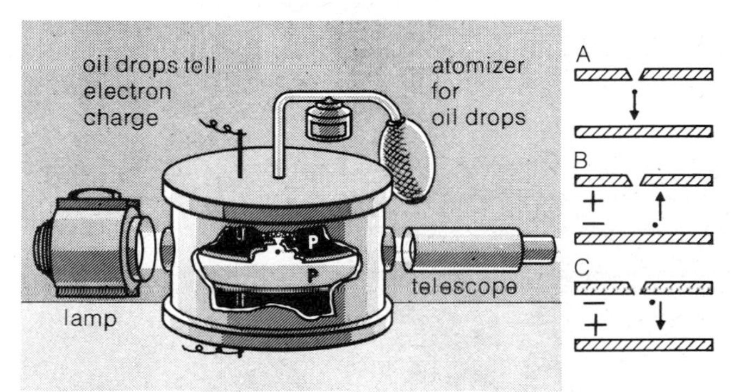

Millikan's oil-drop experiment measured the charge on one particle, and thus told its mass. Tiny oil droplets suspended in air were introduced into the top of an apparatus and observed as they travelled between two plates (p) in the absence of any electron charge (A). From measurements of its movements Millikan was able to calculate the mass of an oil drop. Next, X rays were used to ionize the air in the chamber, and opposite charges were placed on the plates. When an ion stuck to the drop, the drop started moving toward the plate with the opposite charge (B). After timing this motion, Millikan reversed the charges (C), timed the resulting motion, and computed the charge on a drop. By using the value of the charge in Thomson's ratio, Millikan was also able to determine the mass of an electron.

of beams and arriving at the detector as a single spot does not seem tenable. Each individual event is one of transfer of energy to the detector, say, the blackening of a photographic plate, at a single fairly localized spot. If the experiment is performed with very few electrons, there will be only a few spots but each of them localized. These spots will only appear where they are allowed by the diffraction pattern. Where the diffraction beam is intense there will be many; where it is weak, few; and where it is canceled out, no spots at all. The picture of a diffracted wave is palpably related to the beam of many electrons, but if one thinks of a single electron the findings do seem strange; in the end, how can it be that the electron, having spread itself out through all the holes in the grid, has arrived at one particular spot?

It is worth noting that this creates the same paradox that arises when one considers light, although from the point of view of classical theory one must start with an entirely different preconceived notion of light. Classically, light is a wave phenomenon, and diffraction and interference experiments are the best evidence of this. The detection of interference is by experiments exactly the same as those for electrons, say, by blackening photographic plates. The actual effect observed on the photographic plate is a series of dots, on each of which a "packet" of light appears to have been concentrated so that it could react with one atom. There is nothing basically different between the black dots produced by electrons and those produced by light, and this led Einstein to his version of photons—particles of light that were quite unthinkable according to classical wave theory. In quantum mechanics, instead of having electron particles on the one hand and light waves on the other, each of these entities seems to combine the characteristics of particle and wave.

This concept of a "wavicle" strains the ordinary imagination, but we might get some help from that eminent British mathematician C. L. Dodgson, who, in 1865, wrote "The Dynamics of a Particle" and, at the same time, in his more famous identity as Lewis Carroll, gave us Alice and the smile of the dematerialized Cheshire cat. Or we might think of a cartoon in which the tracks of a skier are on a collision course with a tree. The left ski track goes round one side of the tree and the right ski track goes round the other, but they merge again for the rest of the run. All one can assume from this evidence is that the body of the skier dematerialized on one side of the tree and rematerialized on the other.

The Uncertainty Principle

The either-or of quantum mechanics describes something, but does it explain anything? Physicists believe that it does in the sense that they have a precise mathematical model that incorporates features of the wave theory and other features of the particle theory. This combination gives them the means of understanding in great detail a vast number of experimental results that would be contradictory in terms of either the particle theory or the wave theory alone. Physicists would say that there are no paradoxes in the mathematics of quantum mechanics; paradoxes appear only when an attempt is made to limit explanation to the restrictions of everyday language. To illustrate, they could try to use the wave notion to the fullest possible extent and to ignore the particle notion as much as possible. They could get a fair way with this. For instance, they could pretend to forget the Bohr-Rutherford model of the atom, proposed ten years before physicists Werner Heisenberg, Erwin Schrödinger, and Paul Dirac in 1925 contributed to evolve quantum mechanics.

Physicists Niels Bohr and Ernest Rutherford imagined an atomic nucleus surrounded by a group of electrons circling around as if they were planets moving around the sun; this was a kind of solar system of particles. Modern physicists, however, describe detailed features of atomic, molecular, and crystal structure much more verifiably as a continual and continuous wave motion existing throughout the space surrounding the nuclei. They can give convincing descriptions of how the vibrations forming the waves are distributed in space. The geometric distribution of these vibrations is the key to shapes and patterns that reveal themselves, for instance, in the detailed structure of crystals. But it does not explain what actually happens when two atoms hit each other and exchange energy and momentum. Instead of giving up the particle description altogether, they save it partially by saying that electrons are particles that are not governed mechanistically by the laws of cause and effect, as demanded by classical physics. Instead, they are governed by probability laws.

These probability laws derive from the Uncertainty Principle formulated by Heisenberg in 1927. He showed that, in the light of this wave-particle duality of matter, precise determination of the position of a particle will be achieved at the expense of accuracy in specifying its velocity, and vice versa. He also showed that it would be logically impossible to determine exactly the energy of a system at any specific instant. The "impossible" is emphatic and is not a question of improving methods; any observational technique itself will impose the uncertainty. Suppose, for example, that the velocity of a particle is known and that it is decided to find its position by shining a light on it. The "punch" packed by the photons of light impinging on the particle will change its momentum, invalidating the observed velocity. It seems to be an inescapable dilemma.

The uncertainty, however, applies rigorously only to an individual particle. Given a "population" of particles, in which the position of one particle and the speed of other particles can be observed, it is possible to say with precision what the probable characteristics and behavior of that *type* of particle will be. The quantum "model" contains more than is involved in ordinary theories of probability such as are used for games of chance, population statistics, or even the description of complicated physical systems in statistical mechanics.

The Heisenberg Uncertainty Principle appears to the majority of physical theorists to define with satisfactory precision just how far all possible observations can be related to models that can be visualized. Einstein, to the end of his days, was one of the doubters, unable to accept the probability description as something final. And there are also those who would seek to prove that the particle aspect is basic and that the wave aspect might be explained in terms of a not fully understood field of force.

The Principle of Complementarity

Bohr refused to accept the "either-or." He argued that wave and particle were just two different ways of looking at the same thing; they were not mutually exclusive. He developed the Principle of Complementarity, according to which both interpretations are accepted as complementary, not contradictory, and are equally valid observations of the same phenomena. This prescription for peaceful coexistence, not of the protagonists but of the phenomena, need not be confined to subatomic physics. The neurologist, the biochemist, and the psychologist all study the brain somewhat differently, but their objective findings are not contradictory; they are complementary.

The Principle of Complementarity, therefore, would appear to give a contemporary relevance to the theories of

Anaxagoras, who, 2,500 years ago and without the benefit of a $100 million particle accelerator or quantum mechanics, had arrived at some interesting notions. His universe—that is, the world and everything on it and around it—was composed of entities infinitesimal in size that were transmuted and reembodied but eternally conserved. He sought to find a unifying principle that would account for the behavior of each and every particle and each and every part of the universe and would govern all phenomena. He assumed the complementarity of the living and the nonliving. He gave to his "uniform substance" inherent, eternal, and organic properties. The extension of this to *nous* was consistent with the Principle of Complementarity—the regularity of the universe was rational, therefore *cosmos* (order) demanded *nous* (reason).

3.
There Should Be No Night—Energy in an Expanding Universe

Olbers' paradox, which led to the theory of the expanding universe, supported by phenomena from spectroscopy such as "the red shift," with explanations of other phenomena, including exploding stars, pulsars, black holes, and quasars

What we know about energy and matter derives in large measure from what we have learned about the nature and behavior of light. If night were not divided from day by the world turning its back away from the sun, we would not see the stars, millions and millions of which are far, far brighter than our parent star. And the phenomenon of darkness provided "Olbers' paradox" of 1826, which is one of the bases of modern cosmology.

The paradox is simple and yet profoundly significant. As anyone can do on a clear night, Wilhelm Olbers, a German astronomer, observed that some stars are very bright, some are not so bright, and some are dim. The easiest assumption to make is that the brightest stars are the nearest, the medium-bright stars farther away, and the dim ones farther away still. One can also intelligently assume that there are stars even more remote that are so faint that they cannot be seen individually. But Olbers asked whether those remote stars might not be so vastly numerous as to floodlight the night sky. He found that in trying to find effects from regions too far away to be seen in detail he had to make assumptions about the depths of the universe. And the assumptions he made were reasonable in terms of what was learned later.

Olbers first assumed that the distant regions of the universe would be consistently like the Earth's immediate celestial neighborhood. He estimated that there would be distant stars with similar radiation and spaced at the same average distance as the stars that can be seen. In other words, Olbers assumed a symmetrical dispersal in all regions of the universe. He accepted as fact that the laws of physics would apply in the uttermost regions as well as in the innermost—for instance, that light spreads out after leaving its source, as the light spreads from a candle. Finally, Olbers made the assumption that the universe was static, so that the fixed stars would be continuously propagating light from their prescribed positions in the heavens. Thus, the universe would consist of a series of layers of stars (like a luminescent onion). If concentric layers were added one to another without limit, the amount of light the Earth would receive from each layer would be the same, regardless of the radius of each layer, and the amount would increase so that an infinite amount of light would be received from layers stretching to infinity.

There is an obvious limiting factor in that there would be shadows arising from intervening stars that obscure those beyond. This would keep the sum of light from adding up to infinity. Nevertheless, the sum, reduced by such subtractions, still gave Olbers an incandescent glow theoretically equal to 50,000 times the light from the sun at its zenith. He tried to account for the common experience that this is not actually so by assuming that the reduction was due to obscuring clouds of nonradiating matter in space. This assumption did not help him because such matter would be absorbing heat until it, too, would radiate secondhand heat and would no longer serve as a screen. Many attempts were made by Olbers and others to escape from this dilemma. His physics, based on the assumption of fixed stars, was inescapable, but his conclusions would have meant that not only would there be no night and day but that we would be subjected to temperatures of about 10,000° F (5,500° C), whereas the highest temperature on Earth is about 125° F (50° C).

Olbers' paradox made men of science reexamine his assumptions. They observed that the stars are not disposed like the spectators in an infinite sports arena. They form galaxies, a series of stellar arenas of which there are billions like our own Milky Way. This alone would not have invalidated Olbers' assumptions because for the stars in his theory one could substitute galaxies and the argument would still hold; the galaxies and the clusters of galaxies would still give incandescence. Nor does his assumption of the here and now of light (whereas we know that light travels with a speed of 186,000 miles per second so that the light he was observing from any star had originated, not on the night he was looking, but millions of years earlier) resolve the dilemma.

No, the root of Olbers' paradox was his assumption of a static universe. Perversely, his theories demonstrated that the universe must be expanding. The facts that the night sky is dark and that we are not incinerated by Olbers' 10,000° F are evidence enough. Why? Because, if the universe is expanding, then the distant stars are moving away from us at high speeds, and a simple truth is that the light emitted by a receding source is reduced in intensity compared with the light emitted from a source at rest. The clusters, galaxies, and individual stars are escaping as fast as they can from Olbers' paradox.

Spectroscopy

The present evidence for the expanding universe lies in the highly refined methods and measurements used in studying what the English philosopher Francis Bacon called "God's first creature, which was light." And we can legitimately define "light" as "all electromagnetic radiation" so that we can include the new evidence of radio astronomy. The methods and measurements are provided by spectroscopy. The simplest example of a spectrum, the phenomenon that the spectroscopists examine, is the rainbow. This is simply the breaking up of the white light of the sun into its component colors (or component wavelengths). A glass prism or a grid system can do the job more precisely. The sunlight can be spread out, like a scroll, and the colors separated into their wavelengths—blue on the left and red on the right. But the colors do not form a smooth, continuous band. In places dark lines run across the spectrum. They are mainly due to the light of the sun shining through the cooler gases of the sun's atmosphere, gases that happen to be opaque to particular colors and thus produce the effect of thin shadow lines.

Modern astronomers have powerful telescopes and spectroscopes of high susceptibility and spectral precision so that they can analyze the light of individual galaxies and stars. The distant galaxies yield little light, and therefore their spectra are not as clear as the spectrum of the sun, but they do show the prominent dark lines. And these lines are not where they would be in the matching spectrum of the sun. They are "shifted." The shift is always toward the red, that is, to the right. The fainter the galaxy, the greater the shift.

The Red Shift

In wave physics the shift toward the red end of the spectrum always indicates the velocity of recession. To understand this, an analogy can be made with that other wave phenomenon, sound. Suppose you are standing in a railway station; an express train is approaching at a high speed, with its whistle blowing. As it passes you, the pitch of the whistle drops suddenly. This has nothing to do with the whistle or with your ear; it is due to the speed of the train. This can be explained as follows: Suppose the velocity of sound to be exactly 344 meters per second. The approaching whistle is 344 meters away. Half a second later it is 172 meters away. Treat these as two separate instants and it will be seen that the second sound reaches you, the stationary listener, only half a second later than the first. Now, instead of considering the sounds one half-second apart, let us say that the high-pressure points of the sound wave are a thousand cycles, so that the peaks of the wave are only a thousandth of a second apart. In that thousandth of a second the train whistle will have moved 0.172 of a meter and the sound after that will take only 0.002 of a second to reach you on the station platform. The wavelengths are being shortened because of the movement of the source and are therefore being compressed to a higher pitch, or shrillness. As the engine passes the process is reversed, and each successive sound wave has a greater distance to travel. The interval between the successive peaks, therefore, is longer than the interval between their emission, and, accordingly, the pitch of the sound will be lowered. This is what is known as the Doppler effect.

The same thing applies to light waves, but in their case the speed is not that of sound, about 380 yards (344 meters) per second, but of light, which is 186,000 miles (300,000 kilometers) per second. The instrument is not the human ear but the spectroscope, spreading out the radiations into their color frequencies. The visual equivalent of an increase in auditory pitch is a shift of the spectral lines toward the violet end of the spectrum, and a decrease of pitch, that is, an increase of wavelength, is equivalent to a shift toward the red end.

A "red shift," therefore, indicates the velocity of recession of the source (such as a star). If the faintness of a galaxy is accepted as an indication of its remoteness and the red shift of the spectra as the velocity of recession, then the velocity of recession is proportional to the distance of the object.

This, as will be seen, is the transfer to remote space of physical experience observed and measured on Earth. It is valid only if there is a uniformity throughout the universe, only if there really is a cosmic order. This has to be assumed before it can be examined. And, as it is examined, the more efficient the means of observation, the more convincing the arguments of uniformity become. If the medley of the heav-

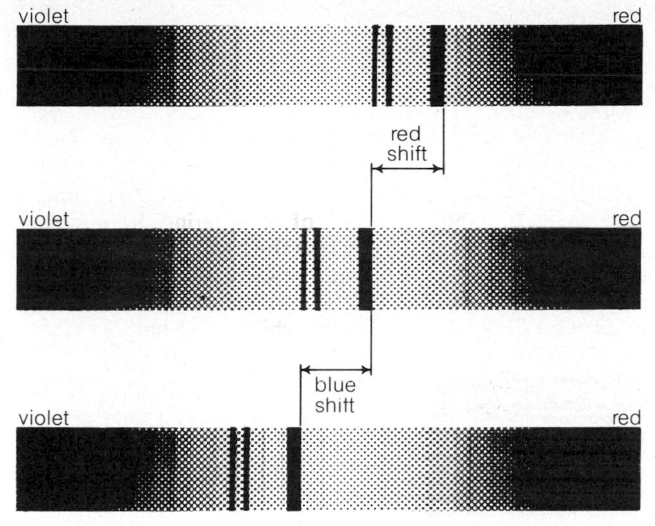

The absorption lines from a star or galaxy shift to longer wavelengths (red shift) when the object is receding from an observer (top). They shift to shorter wavelengths (blue shift) when the object is approaching an observer (bottom). The amount of shift depends on the velocity of the object in relationship to the observer: the greater the velocity, the greater the shift. The absorption lines from the sun (center) are used for comparison.

ens is sorted out and one looks closely enough (in detail) and far enough (in distance), remarkably consistent patterns emerge and repeat themselves. It is like the design of a wallpaper: the details within a pattern may vary, but the pattern is reproduced consistently and is the same in all directions.

If by rigorous and repeated observations one is satisfied that there is a general pattern, and when each new discovery is consistent as a detail in that pattern, then one is entitled to assume that this is one of the uniformities of nature. It is also permissible to assume that that pattern will appear the same from every vantage point, whether it be that of an invalid lying in bed studying the cosmic wallpaper, an intelligent fly on one of the galactic designs on that wallpaper, an observer at Mount Palomar looking at the universe through a powerful telescope, or his "opposite number" on a planetary world in another island universe looking through a similar telescope.

Pushing Back the Galaxies

In 1923 the U.S. astronomer Edwin Hubble looked at this pattern in a particular way. At Mount Wilson Observatory he was studying the spiral nebula of Andromeda. A spiral nebula is a characteristic feature of the repetitive pattern; it has a center shaped like a lozenge with spiral arms winding around it. Until Hubble had a special look at Andromeda, these nebulae were supposed to be located among the stars of the Milky Way. He noticed that the spiral arms of Andromeda contain a number of extremely faint stars, the brightness of which changes periodically. Such stars are called Cepheid variables. Their periodic changes in luminosity are explained by the periodic pulsations of their giant bodies. The brighter the star, it has been observed, the longer the period of pulsation.

Harlow Shapley, a Harvard University astronomer, used those pulsations as a means of measuring distance; by measuring the pulsations he could establish a star's absolute brightness and this, in measured contrast to its visible brightness, gave him the actual distance to such a star. Using this method, Hubble assured himself that many stars in Andromeda had high absolute, initial luminosity and yet were so

Andromeda Nebula, the great spiral galaxy in the constellation Andromeda, is visible to the naked eye. It was the first galaxy proven to be located beyond the Milky Way.

faint visually as to be at the limit of visibility. (This is like saying that a powerful searchlight showed up as feebly as a match.) If this were so, then there was only one possible explanation—that they must be very, very far away. By systematic measurements of this kind it was shown that Andromeda is not a lodger in the Milky Way but a galaxy in its own right, at least a million light-years away.

Walter Baade, also of Mount Wilson Observatory, was able to resolve photographically the central body of Andromeda and the associated swirls and to sort out individual stars. It is now known that Andromeda and many, many more spiral nebulae each consist of billions of stars like, or greater than, our own sun. What had once looked like luminous sandstorms have emerged as island universes. The individual stars blur together into a faintly glowing mass only because of their distance from the observer.

Hubble's discovery was followed up by the spectroscopists. The light emitted from the spiral nebulae showed a shift to the red; that meant, in terms of the Doppler effect, that they were moving outward, ever outward, like stellar wagon trains pushing out to new frontiers. In the diorama of the universe those nebulae that had once seemed part of the foreground of the pattern of the Milky Way were seen to be at remote depths. Scientists now recognize that the entire space of the universe is populated by billions of galaxies—not only the upgraded "spiral nebulae" but also elliptical and spherical galaxies—each with billions of stars that almost certainly have their own solar systems of planets. What is more, all of the galaxies and stars composing them are flying away from each other at fantastic speeds.

The notion of an expanding universe is physically and mathematically unavoidable if we are to escape from Olbers' paradox (and incineration by his incandescent fixed stars) and to account for the red shift. To express it as a mathematician would, the universe has uniformity. But this leads to another question: how can it move and still maintain its uniformity? The answer is that it can only move in such a way that the velocity of every object is in the line of sight (the line joining the Earth and a distant astronomical body) and proportional to its distance. This is the only type of motion that will maintain uniformity. An expansion with the velocity of recession proportional to distance, therefore, is a natural consequence of the assumption of uniformity, which is based on confirmed observation. Any theory of the universe, consistent with the observed facts, must always come up with the answer that it must be in motion with objects showing velocities proportional to their distances.

Taking further liberties with the pattern of the universe, let us, instead of hanging it on the walls, put it on a child's balloon. Just as the Declaration of Independence can be photographed down to the size of a printer's period or the "microdot" of the spy thrillers, so can all the clusters, galaxies, and stars, be printed on the balloon. (The sun would be just an atom of carbon in the printer's ink.) As one proceeds to blow up the balloon, the congestion begins to sort itself out, and each item in the design moves farther and farther away from the other items. The galaxies become differentiated in the clusters, and the stars in the galaxies become separate specks. What this model shows is the recession caused by the uniform expansion of the entire system; what it cannot show is that galaxies also possess an individual random motion similar to the thermal motion of molecules in a gas. The two kinds of motion interact, and sometimes, if the random motion is toward the observer, the spectrum may show a violet shift (to the left) instead of a red shift (to the right). This might seem to belie the general argument of expansion but does not because at greater distances the increasing recession velocities are too great to be affected by the random "thermal" velocities. Thus the expansion of the system as a whole is not challenged.

The Expanding Universe

The expanding universe is generally accepted. Astronomical observations provide more and more supporting evidence, and the scientific world accepts it. Anyone who disagreed would be considered as eccentric as a Flat-Earther.

The checking, however, continues. The cosmological theorist says to the practicing astronomer, "Look out for so-and-so," and starts a new chain of inquiry. Or, as has often happened, the radio astronomer will say to the optical astronomer, "I am getting powerful radio signals from a source that does not appear on your star charts. See if your telescope can locate it." Two of the most powerful radio sources, for example, are on the bearings of the constellations Cassiopeia and Cygnus.

At the Palomar Observatory, Baade used the telescope to examine the Cygnus region and found that the radio source was visible. It was 500 million light-years away. The distance and the power of the signals showed that this was indeed a remarkable phenomenon. Various explanations have been offered for "Cygnus A," this transmitter operating with a strength of 10,000,000,000,000,000,000,000,000,000,000,000 kilowatts. One is that a star exploded and triggered a chain reaction of other star explosions, causing the production of electrons traveling with nearly the speed of light and causing intense radio waves. Another theory (now generally discarded) is that it was a collision of two galaxies. Another is that it was the birth of a galaxy. In this process the gas cloud (the protogalaxy) contracts and breaks up into smaller clouds that become more dense and form the stars. As this happens, gravitational energy is released, which generates cosmic rays. These collide with gas atoms and discharge the great amounts of high-speed electrons necessary for intense radio emission.

In 1974 U.S. astronomers Roger Blandford and Martin Rees suggested that the visible star in Cygnus was a spinning mass of gas, a strong source of energy that was ejecting narrow jets of fast-moving particles. The jets would emerge in exactly opposite directions along the axis around which the clouds of gas were spinning. These jets would produce a

"Cygnus A," located in the northern constellation Cygnus, is one of the most powerful sources of radio waves, though the source of the energy emitted remains unknown.

shock wave in the thin gas of space. This shock wave would generate the radio noise being received on the Earth as though it were coming from twin transmitters on either side of the visible star.

Exploding Stars

Radio astronomers, heirs to the wartime radar experts who refined the methods of microwave detection, register the birth cries and death gasps of stars. In the here and now they can carry out postmortems on stars that died millions of light-years ago, and in the black holes of space they can discover where the corpses are buried.

Three stars have been seen to explode in our own galaxy and to the naked eyes of three incontrovertible witnesses. The first witness was Yang Wei-te, the calendar-keeper of the Sung emperor of Khaifeng. In the summer of A.D. 1054 (our calendar, not his) he reported that a new star had appeared in what we call the constellation Taurus. As bright as the planet Venus, it was visible even in daylight. It was reddish-white in color. Within a month it had gone. The second witness was the Danish astronomer Tycho Brahe, who ob-

The aperture-synthesis radio telescope at Stanford University in California is composed of five antennas located on an east-west baseline.

served an occurrence in 1572, and the third was the German astronomer Johannes Kepler in 1604.

When radar scientists went back to academic life after the war, they turned their improved radio detectors onto the heavens, first with the modest intention of tracking meteors as they had once tracked aircraft. When they turned their mobile receivers to scan the firmament, they got unexpected signals. These were coming from the stars themselves and particularly strongly from a locality in the Crab Nebula, in the direction of the constellation of Taurus. Nine hundred years after Yang Wei-te had actually seen the new star, observers were getting a radio commentary from it.

Today the debris of that star is plainly visible with a large telescope. It is the Crab Nebula, a luminous shell of gas expanding in all directions. It is glowing white with red tassels of hydrogen swirling from it. Within it are strong sources of radio and X rays. It also contains a pulsar.

Pulsars, Black Holes, and Quasars

Pulsars were identified at Cambridge University in 1968. They

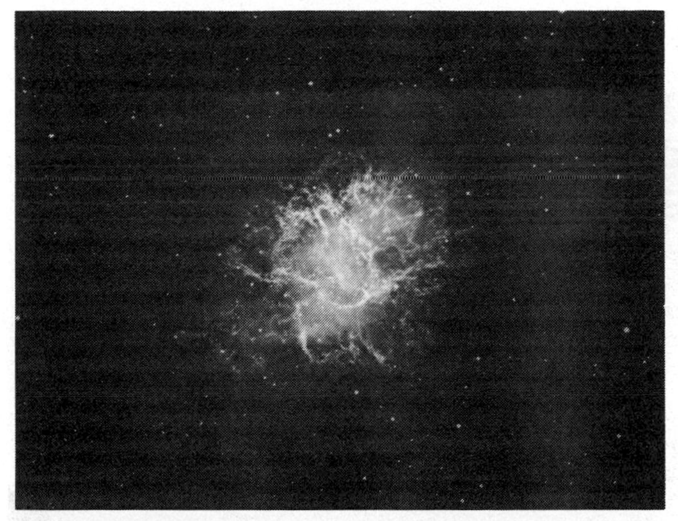

The Crab Nebula is assumed to be the remnant of an exploding star that was observed by Chinese astronomers in A.D. *1054. Modern astronomers have detected radio waves, X rays, and a pulsar within the Crab.*

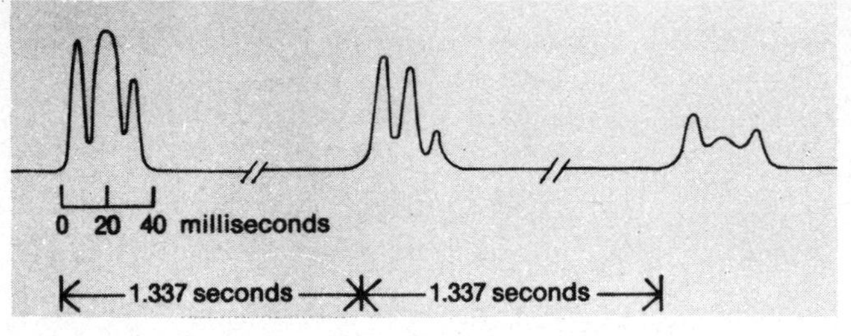

A series of radio waves indicate the three-pulse pattern emitted from CP1919, the first pulsar detected. The individual pulsations vary considerably in amplitude, but the repetition rate remains constant.

are "pulsating radio sources" that pump out bursts of radio energy with the regularity of a ticking clock. The one in the Crab Nebula was precisely pinpointed by radio, after which the optical astronomers turned their telescopes onto it. It was visible, and since then many pulsars have been located by radio and then seen. They are characteristically blue, emitting ultraviolet light as well as visible light. They seem to be associated, wherever they are, with the remains of exploded stars called supernovas. They also appear to be small and of dense mass. However compact, they must encapsulate an enormous amount of energy because they are pumping jets of it many times a minute with absolute regularity.

Astronomers were already familiar with "white dwarfs," stars once gaseous and of large dimensions that ran out of energy. After a series of enlargements and puffs they shrank and became solid bodies. A star with the dimensions of the sun would shrink to the size of the Earth, only be much denser. The original theory was that the forces of gravity had overcome the electromagnetic forces and compressed the nuclear forces. Thus, the electrons would be squeezed into the protons to make neutrons. In this theory a star with a mass greater than that of the sun would be crushed into a ball only about twelve to eighteen miles in diameter—a neutron star.

The pulsar might be a version of such a neutron star. But even before it was discovered, there was a school of thought that went even farther. The white dwarf, the pulsar, or the neutron star might just be stages in the collapse of a star crushed by the forces of gravity. It could be calculated that if a star more massive than the type that produces (and stops at) the neutron star were to collapse, gravity would overwhelm the nuclear forces as well as crush even the neutrons. Gravity would increase until all radiation and all light would be extinguished. Such a star would become a black hole. The star would have vanished but not without a trace because there would be the ring of intense gravity where it had been. According to some the black hole, the celestial drainpipe, would be the cosmic waste disposal unit because everything that went into it would totally dematerialize at the point of "singularity," where all existence and time itself would vanish.

The Marquis de Laplace, an eighteenth-century French astronomer and mathematician, had reasoned that light, having mass, would be subject to the force of gravity. If a star were massive enough, gravity would prevent light from escaping into space. Gravity would enclose it like the shutters of a dark lantern. That could reconcile latter-day scientists to the presence, in space, of gravitational locations that should have been associated with a massive body. By Laplace's definition, these were black stars. The remorseless logic of the black hole, however, goes much farther: it pursues matter into a physical nirvana or total oblivion. All that is left, like the smile of Lewis Carroll's Cheshire cat, is a measurable gravitational force where the vanished star once was.

Speculation about the behavior of energy in the universe must take into account the observed characteristics of the quasars. These are quasi-stellar objects with tiny, ill-defined optical images. Some exhibit radiation at radio wavelengths, but more often they are characterized by emissions in the ultraviolet part of the spectrum that are much greater than in normal stars. The red shift in the spectrum is exaggerated beyond any found in the spectrum of ordinary galaxies. The first suggestion was that the quasars must be very far away and many times brighter than the brightest galaxies and that they are slingshots of energy catapulted by galaxies until they are traveling with a velocity close to the speed of light. This view was questioned, however. The red shift might not be due to the receding distances of the expanding universe. It might be caused by strong gravitational red shifts in galactic systems not so far away, which can be ascribed to atomic transitions. It has been argued that they can be generated in multiple supernova outbursts or in collisions between galaxies or in the collapse of a supermassive star. It is easier to find quasars than to explain them. If they are associated with the collapse of supernovas, they may be roadmen's lamps around that demolition site, the black hole.

4.
The Sun—
The Source of All Energy

Visible features of the sun, including sunspots, prominences, and the corona; spectroscopic analysis of the sun's rays and what it shows about the sun's magnetic fields, motions, and structure; and the sun's source and output of energy

The sun is a middle-sized, middle-aged, middle-class star, existing in an outer suburb of the Milky Way. Its nearest star neighbor in the galaxy is Proxima Centauri, more than four light-years away. The Earth is just over eight minutes away, which means that light leaving the sun and traveling at 186,-000 miles (300,000 kilometers) a second takes eight minutes to reach it. In terrestrial measurements the sun is 92,957,000 miles (149,600,000 kilometers) away from the Earth. Its diameter is 109 times and its mass is 333,000 times those of the Earth. Its mean density is 1.41 that of water. The density actually varies from a value in its outer layers much less than that of air to about sixty times the density of the Earth's rocks at its center.

The computerized versions of various models of the sun's characteristics provide pretty fair agreement about its life span—something like 100,000,000,000 years. A plausible version of the sun's development, derived from the Big Bang theory, would be as follows: for its first million years the sun's share of the swirling gases was (in Isaac Newton's term) "convening," and a radioactive core was forming and steadily growing to include nearly all of the sun's mass; at 23 million years internal nuclear reactions began; at 26 million years the radioactive core was well established, a self-contained furnace; and by 37 million years the sun appeared much as it is today.

The sun, since it always appears circular in the sky, must be spherical, and its bright visible surface is called the photosphere. As we now observe it, the brightness of the photosphere is not uniform; the region near the circumference, or limb, is appreciably less bright than the center—a fact that appears most obvious in photographs.

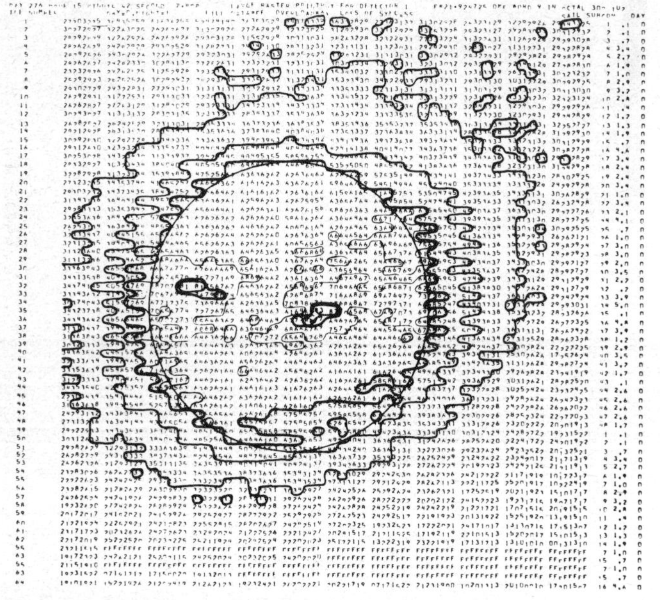

A computer-generated image of the sun, transmitted to Earth from an orbiting satellite, was recorded two hours after a solar flare occurred in the region near the center of the sun's disk. The smooth circle depicts the approximate size and position of the visible sun; the wiggly lines are isophotes, separating regions of different ultraviolet intensity.

Sunspots, Prominences, and the Corona

Sunspots, which are frequently visible to the eye when viewed through a densely smoked glass, are the most easily observed markings on the sun. A sunspot consists of a central, apparently black area called the umbra, surrounded by the less dark penumbra. Each of these regions is in fact exceedingly bright and appears dark only by contrast with the still brighter photosphere. Spots often are clustered together in groups and only rarely appear outside two zones on either side of the solar equator, extending about halfway toward the sun's north and south poles of rotation. The spots move steadily across the face of the sun so as to leave no doubt that the motion is caused by the rotation of the sun about an axis.

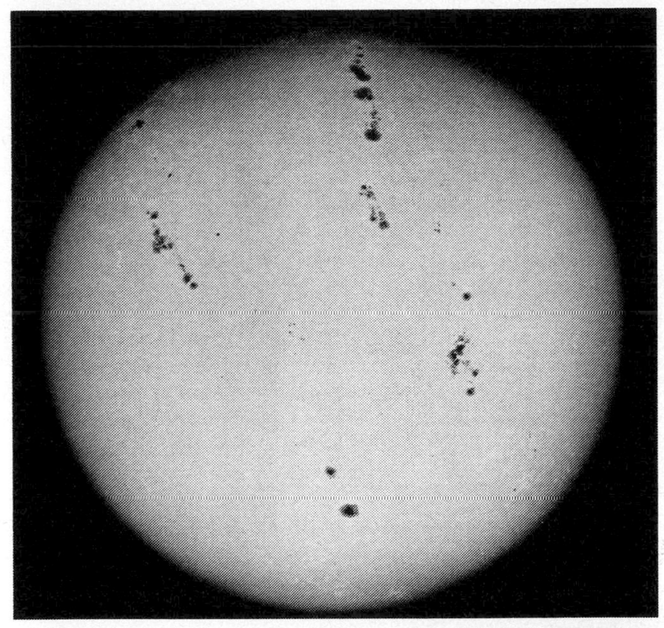

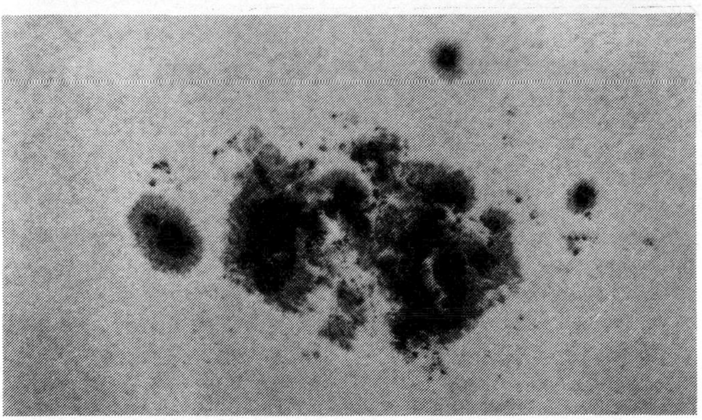

Clusters of sunspots are revealed in a telescopic view of the sun (top) taken at a sunspot maximum on December 21, 1957, but even between such peak cycles of activity sunspots—such as this large and complex group photographed on May 17, 1951 (bottom)—do occur.

A remarkable feature of the sun's rotation is its variation with latitude; that is, the sun does not rotate as a rigid body: the nearer to its equator, the faster the rotation. Near the equator the mean rotation period is 24.65 days; at solar latitude 20° it is 25.19 days; at 35°, 26.63 days; and at 60°, 30.93 days.

Individual sunspots appear spasmodically, remain visible for periods varying from a few hours to several months, and then disappear. This apparently capricious behavior becomes a striking regularity, the annual number reaching a maximum approximately every eleventh year. A similar regularity characterizes the location of the spots. At a time of minimum sunspots those of a new cycle begin to appear in the higher latitudes, both north and south, of their appointed belts, and as the cycle progresses the place of outbreak gradually moves toward the equator.

During the comparatively rare occasions on which the sun is observed in total eclipse, small red clouds are seen apparently floating above various points on the circumference of the dark moon. They belong in reality to the sun and are known as prominences. Prominences are not seen through an ordinary telescope in full daylight because the intense photospheric light, diffused by the Earth's atmosphere and the optical parts of the telescope as by a ground glass, acts as a veil through which prominences and faint stars alike are invisible as individual objects. Prominences assume various shapes and sizes, sometimes reaching heights of hundreds of thousands of miles above the photosphere. The larger prominences can easily be seen during eclipses.

Prominences seem to be extensions of an outer solar atmosphere that is directly revealed at the beginning and end of each total eclipse. When the moon is just covering the last remnant of the photosphere, a crescent of atmosphere remains exposed for a few seconds. It is easily seen in a small telescope as a brilliant red layer and has been named the chromosphere.

The most striking solar eclipse phenomenon is the corona, a pearly white halo enveloping the sun and extending in more or less definite rays or streamers to distances of many solar radii. The brilliance of the corona diminishes rapidly with distance from the sun's limb, and although its total brightness is about one-half that of the full moon, it is much fainter, area for area, than the solar prominences. No two aspects of the corona seen at different eclipses are identical; however, there

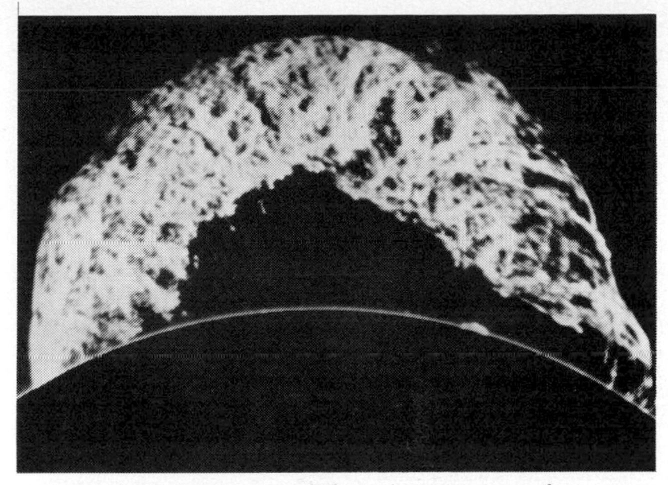

A large solar prominence erupts from the sun at a speed of 160 kilometers (100 miles) per second. This huge surge of incandescent ionized gas has a length of about 640,000 kilometers (400,000 miles).

is a relation between the shape of the corona and the sunspot period. At sunspot maximum the corona appears to extend equally in all directions from the sun's limb and appears

The silvery halo of the solar corona was clearly visible during a total eclipse that occurred on November 12, 1966.

round. At sunspot minimum the poles of the sun are marked by comparatively small tufts of light, while from the equatorial regions long streamers extend to enormous distances.

A special type of telescope, a coronagraph, can be constructed so that it diffuses almost no light and intercepts the image of the photosphere with a metal disk. On high mountaintops, where diffusion by the Earth's atmosphere is negligible on days of exceptional clarity, the innermost, brightest portions of the corona can be observed directly with this instrument.

Analyzing the Rays

The spectroscope is the most important of the auxiliary instruments used with the telescope for the investigation of the sun because it can select minute portions of the flood of solar radiation for intensive study. Nearly all solar research depends at some point on a device for sorting radiation by its wavelength. This sorting results in the familiar sequence of colors seen in the rainbow. The word *spectroscope* in its most general meaning is used to indicate such a device. The human eye is sensitive to only a small band out of the vast range of wavelengths represented in the total radiation of the sun, but over this restricted range the word *wavelength* is analogous to a precise specification of color. The spectrum (the ordered arrangement of radiation obtained from a spectroscope) of the sun's radiation consists mainly of a continuous bright background on which dark lines appear. This absorption spectrum confirms the direct observation during an eclipse that the photosphere is covered by a layer of glowing gases. The dark lines are evidence that the glowing gases are relatively cool.

A qualitative spectroscopic chemical analysis of the sun's outer envelopes may be made by comparing the positions, or wavelengths, of the absorption lines with those of emission lines produced by known substances in the laboratory. In this manner the presence on the sun of sixty-seven elements known on Earth, and eighteen chemical compounds, has been established. The spectroscopic chemical analysis of the sun can also be made quantitative by directing attention to the total amount of radiation extracted from the continuous background by the dark lines. The darkness of a line is related in a complicated way to the amount of material that produces it, but with the aid of laboratory calibrations of line strengths the relative abundance of the chemical elements has been determined for the sun. The chemical constitution of the sun resembles that of the Earth except that the lighter elements, especially hydrogen and helium, account for almost all of its mass.

The spectrum of individual features of the sun may be studied, as well as the light of the entire solar disk. Just at the start of, or just before the end of, the total phase of a solar eclipse, when the chromosphere still remains exposed for a few seconds, the spectrum of its light shows bright rather than dark lines. This occurs because the chromosphere is semitransparant, and there is no bright photospheric background to provide a source of radiation to be absorbed. Because it appears so briefly this spectrum is known as the flash spectrum, and it is essentially identical with the absorption spectrum in the position of its lines. The heights reached by the various substances have been determined from special observations at such times. The majority are confined to the lowest atmospheric layers, not more than 500 miles (800 kilometers) high, near the top of the photosphere or the bottom of the chromosphere. Hydrogen, helium, and ionized calcium can be traced to great heights, extending up to 8,700 miles (14,000 kilometers). The radiation of hydrogen is the source of the intense red color of the chromosphere.

The chromosphere is so bright when viewed during a solar eclipse that it seems reasonable to expect success in the observation of its bright-line spectrum in full daylight. Such observation is difficult, however, because the photospheric light diffused by ordinary telescopes and the Earth's atmosphere spreads over and drowns the chromospheric radiations. Nevertheless, the hydrogen and helium emission lines are easily observable with very modest telescopic and spectroscopic equipment, although the fainter lines can be seen only with powerful instruments fed by clean (nondiffusing) telescopes, such as the coronagraph, that are located where the scattering of light by the Earth's atmosphere is as small as possible.

When the spectroscope is directed toward the corona, at eclipse time or with the aid of a coronagraph, it is found that a great deal of the coronal radiation yields the ordinary solar spectrum and seems to be only reflected sunlight. Actually, radiation from the corona contains three main components: a continuous spectrum, which is sunlight diffracted by interplanetary dust particles; a series of bright emission lines, which have not yet been produced in terrestrial laboratories; and radiation that can be detected using radio methods. The bright emission lines of the corona can be identified with radiations emitted by isolated and highly ionized atoms of iron, nickel, and calcium.

Applications of Spectroscopic Analyses

The spectrum of the solar corona is one of the best examples of the results of physical conditions in modifying the radiation emitted from various regions of the sun. After the effects on line spectra of temperatures, pressures, magnetic and electric fields, and motions of the sources of light had been established by means of terrestrial spectroscopic studies, these and other physical parameters of the sun could be investigated by detailed investigations of the solar spectrum lines. These investigations have produced knowledge of magnetic fields and motions of the sun.

Magnetic Fields. Many years of observation have shown that all sunspots are magnets, some presenting a north and others a south pole to the surrounding space. In each pair of spots, or spot groups, the leaders and followers in the journey around the sun's axis have opposite polarities. Many apparently single spots have invisible companions that can be found by the magnetic splitting of the spectrum lines. There is a reversal of polarity in the southern solar hemisphere with respect to the northern; thus, if the leader of a pair were a north pole in the northern hemisphere, it would be a south pole in the southern hemisphere and vice versa.

More refined investigations have shown that weak magnetic fields are widespread over the solar surface. At times, these fields, which are much weaker than those of the spots, give the impression that the sun as a whole is a magnet like the Earth. Continued observation of the sun's weak fields, however, has shown that they vary in a complicated, quite disorganized fashion and produce no resultant general magnetic field for the sun. There is no obvious connection between brightness variations, motions, and the general background of weak fields in the photosphere.

Relatively rare measurements of magnetic fields in the chromosphere indicate a weak correlation between magnetic fields and motions in this layer of the solar atmosphere. A fairly close relationship between brightness, motions, and magnetic fields may exist in the corona, but the technical difficulties of achieving direct, simultaneous observation of these quantities are enormous.

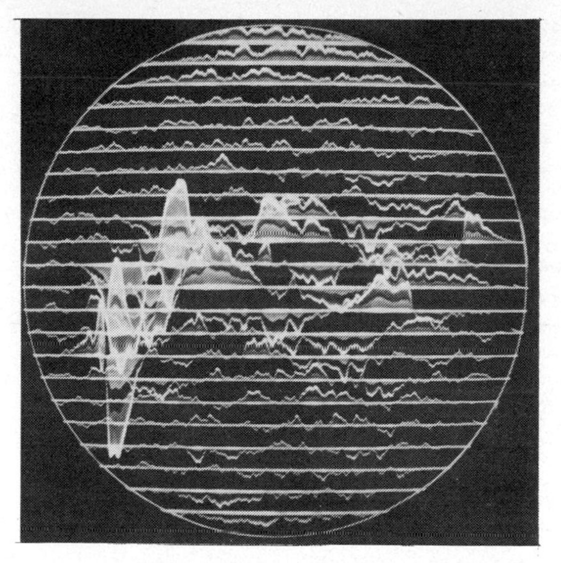

The location, field intensity, and polarity of weak magnetic fields in the sun's photosphere, excluding sunspots, are recorded on solar magnetic maps, which are generated automatically by a scanning system consisting of a polarizing analyzer, a powerful spectrograph, and a sensitive photoelectric detector.

Motions of the Sun. Spectroscopic measurement of line shifts in stellar spectra has been used to find the speed of the sun's motion among the stars. Analysis of the line displacements shows that the solar system is moving at a speed of 12.5 miles (20 kilometers) per second with respect to the local stellar system.

Another application of spectroscopic analysis to the sun consists of the observation of line shifts in spectra of the east and west limbs of the sun. These show that the east limb is approaching the Earth and the west limb receding. The deduced values of the speed of rotation agree very well with those indicated by the spots, and the spectroscopic observations can be extended beyond the narrow belts to which the spots are confined. Similar spectra can be used for determining which lines in the solar spectrum properly belong to the sun, and which are produced by absorption in the Earth's atmosphere. The atmosphere of the Earth contains much water vapor and other constituents that produce many absorption lines. These are not only intermingled with the solar lines but also crowded so closely together in long ranges of the spectrum that no sunlight at all reaches the Earth in these wavelengths. Such lines, however, occupy identical positions in the spectra of the two solar limbs and are therefore readily distinguished from the displaced lines of the sun.

One of the most difficult and important applications of the shifts arising from motions along the line of sight is the effort to detect atmospheric currents in the sun. Solar storms sometimes occur, all of incomparably greater fury than the hurricanes and tornadoes of the Earth but quite inconsequential as far as the sun is concerned, and the stormy motions of the solar atmosphere are revealed by distortions and displacements of the spectrum lines. Prominences are sometimes ejected from the sun as a result of such activity, and bits of the solar atmosphere can be projected from the sun to Earth, where they may produce auroras and magnetic storms. There are, however, systematic movements that have been observed, and of these perhaps the most interesting are those occuring near sunspots. At the lower atmospheric levels gases move upward and outward from a spot, while in the higher levels the movements are inward and downward, as if a spot were a sort of whirlpool into which the high-level gases are drawn. Indeed, there is definite evidence of high-level circulatory movements around the axes of the spots, which points strongly to the same conclusion.

Away from the spots, the solar atmosphere is in a constant random, bubbling motion in rather strong contrast to the organized movement enforced by a sunspot region. Close to sunspots and their surrounding bright regions (faculae) the bubbling, or twinkling, motions of small regions in the solar surface are strongly suppressed, possibly because of the action of magnetic fields.

Solar flares—the sudden, short-lived brightenings of small parts of the solar surface—are the most significant manifestations from the standpoint of the direct effects of the sun on Earth. The outbreak of a flare, visible in hydrogen light and accompanied exactly at the start of the visible disturbance by a sudden large increase in the solar radiation in the radio region of the spectrum, often is followed by a magnetic storm on the Earth after a delay of one to three days. Magnetic storms are accompanied by auroral displays, and it is assumed

An active region near the edge, or limb, of the sun—as recorded from space in ultraviolet light—shows projecting looplike forms indicative of strong magnetic fields.

that the flares in some way eject nuclei of hydrogen atoms (or protons) and electrons and that these, traveling at rather modest speeds of from 375 to 550 miles (600 to 900 kilometers) per second, impinge upon the Earth's atmosphere about fifty hours after the flare outbreak to produce the observed terrestrial disturbances. The connection between solar flares and the electrical equilibrium of the Earth's ionosphere has been established so firmly that observations of the flares are important elements in the prediction of favorable or unfavorable conditions for the sending of long-distance radio messages.

Structure of the Sun

A mathematical model of the structure of the sun can be based on astronomical data and on data from terrestrial physics laboratories dealing with the behavior of matter at temperatures, pressures, and densities that range from the extremely high to the almost vanishingly small. According to these models, the temperature at the center of the sun is about 20,000,000° K. At the farthest edge of the photosphere it is about 4,200° K. In the chromosphere the temperature starts to rise, at first very slowly and then more rapidly, until at the beginning of the corona, about 9,000 miles (14,400 kilometers) beyond the photosphere, the temperature is 500,000° K. In the outer parts of the corona, 1,500,000 miles (2,400,000 kilometers) from the center of the sun, the radio observations are best fitted by a temperature of 1,500,000° K. Pressures start at 6,000,000,000,000 pounds per square inch (psi) at the center, decrease to about 32 psi near the visible surface, and become inappreciable in the upper chromosphere and corona. The mathematical model indicates that 99% of the mass of the sun is concentrated within 0.6% of its radius; thus, the density of the sun decreases rapidly away from the center.

Only the most elemental forms of matter can exist at the temperatures encountered near the center of the sun. Atomic nuclei must be stripped almost bare of their satellite electrons, and the rates of motion of the elementary particles must be in the range of from 60 to 1,000 miles (100 to 1,600 kilometers) per second. All of the bits of matter—electrons, neutrons, and protons—that exist at the sun's center participate in a process of continual interchange of energy between radiation and motion of particles. But there is an important difference in the behavior of the atoms and electrons on one hand and the radiation on the other. The motions of the particles are kept within a limited range by the gravitational equilibrium of the whole sun, while electrostatic forces preserve a constant proportion between the number of nuclei and the number of electrons in each region. Radiation, however, is not so controlled. It works its way from the center, where it is most intense, out to the surface and thence to space at the observed rate of 3.8×10^{33} ergs per second. (An erg is the amount of work required to raise 1/981 of a gram vertically through one centimeter.) The sun loses radiation at this prodigious rate and has been doing so for many millions of years, while the electrons and atomic nuclei remain chained within its boundaries.

It has been realized for many years that the observed rate of energy loss by leaking of radiation from the solar surface must be balanced rather exactly by the rate of energy generation in the interior, since there is little evidence that the sun is heating or cooling rapidly. Because other sources of energy are insufficient, scientists have sought a series of nuclear transformations that would liberate radiation at the required rate. The transformation of 1% of the sun's mass from hydrogen to helium would supply enough energy to keep it shining for fully one billion years. There are certain conditions, however, that must be satisfied in the selection of possible energy sources. The sun must not only be in balance as regards the flow of radiation but also must simultaneously be in mechanical equilibrium. This means that the pressure, which is primarily determined by the temperature, must exactly balance the weight of the overlying layers at every point throughout the sun. Restrictions such as these on the possible choices have directly led to the conclusion that fusion between hydrogen nuclei, or protons, to produce helium is the source of the sun's energy.

The Sun's Energy Output

Of the sun's output of 3.8×10^{33} (10 followed by 32 zeros) ergs per second, the sun's planets and their satellites (such as the Earth's moon) receive about one part in 120 million, and the rest goes into space. Of that one part the Earth receives an insignificant pittance, but it is on that pittance that all life depends. The energy reaches the Earth in the form of X rays, ultraviolet radiations, visible light, infrared radiations, and radio emissions.

Since the sun is visible, energy is flowing out of it and it cannot be uniform in temperature. The effective temperature must come from the photosphere. This is calculated to be about 9,975° F.

In the extreme ultraviolet wavelength (less than 0.00015 mm) the continuous background of the sun's spectrum is undetectable, and only bright lines appear. The principal line is the fundamental line of hydrogen. The character of this line indicates that it is formed in a region of the chromosphere where the temperature is increasing rapidly with height above the photosphere. Other bright, high-temperature lines are present to support this deduction. Thus, the chromosphere, which is the source of the bright lines in the extreme ultraviolet part of the solar spectrum, is a most unusual region in which the temperature increases with distance away from the sun, and atoms are observed in motion at unexpectedly high speeds.

The result derived from the observations in the extreme ultraviolet spectrum is confirmed by measurements of the sun's radio radiation. In the ultraviolet part of the solar spectrum the wavelengths of solar radiation are less than 0.00015 mm; in the radio region the wavelengths are in the range of 5 mm to 10,000 mm. The shortest wavelengths originate near the base of the chromosphere, but the longest can only come from far out in the corona. In order to calculate an apparent temperature for the sun, energy observed at the various wavelengths can be equated to the energy radiated from a black body of the same size as the sun's visible disk. (A black body is a body or surface that completely absorbs all energy falling upon it.) Apparent temperatures, based on measurements of the radio energy received from the sun, range from about 6,400° K for the shortest waves originating in the chromosphere near the sun's visible surface to 1,200,000° K for wavelengths of tens of yards originating entirely in the solar corona, more than one million miles above the visible surface.

On the Earth lightning flashes are the only natural sources of radio waves, but there are at least five different kinds of radio energy generated on the sun. These include: (1) the continuous, nearly invariable component from which the apparent temperatures are computed; (2) the continuous and steady contribution that is associated with the presence of luminous clouds on the sun's visible surface; (3) erratic increases in energy of extremely short duration that follow closely upon one another, often so closely that they overlap; (4) sudden, great, nonrecurrent increases beginning at the start of solar flares and associated with them; and (5) occasional isolated increases of very short duration. Sometimes the variable portion of the sun's radio energy quite overwhelms the continuous background. This is completely different from the situation in the visible spectrum, where the variable features are difficult to detect.

The Sun's Energy Source

The energy source responsible for the radiation of the sun and stars is a main preoccupation of nuclear physicists. Much of what is now known about energy processes in the sun derives from the work in the twentieth century on the nucleus of the atom. Basic to those studies is thermonuclear fusion, first reproduced on the Earth in the cataclysmic form of the hydrogen bomb.

Nuclei become fused, or welded together, when temperatures are sufficiently high. The heat of the interior of the sun is 20,000,000° C. Even higher orders of temperature are produced at the instant of the explosion of a man-made fission bomb. The percussion cap or detonator of the hydrogen bomb is surrounded by a wadding of material rich in deuterium (double-hydrogen) or tritium (triple-hydrogen),

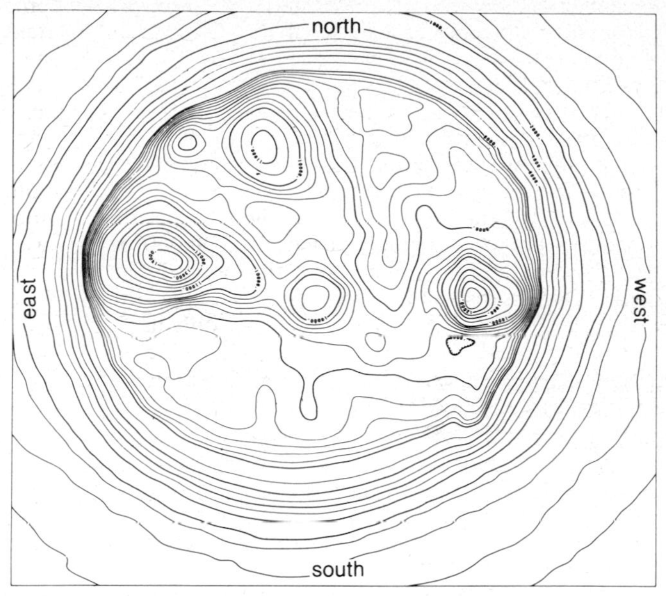

Observations made on a 140-foot radio telescope are recorded on a radio map of the sun taken on March 7, 1970, right after an eclipse.

which can combine in a quick buildup to the combination of four hydrogen atoms to make helium. The excess energy is violently released. In this explosive process man has taken shortcuts to get results in split seconds. The sun does the same thing in a more leisurely manner (it has billions of years).

In the sun two protons combine to form a deuteron, emitting a positron (positive electron) and a neutrino (a particle with zero rest mass and ½ spin). The deuteron captures a proton to form an isotope of helium (^{3}He). Two helium isotopes then combine to form ordinary helium, discarding two excess protons. Three plus three (^{3}He plus ^{3}He) make six; subtracting two leaves four (^{4}He), which is stable helium.

By means of spectroscopic analysis astrophysicists know how much hydrogen is available in the sun (or any other observed star), and they know the rate at which the hydrogen is being consumed by conversion into helium. Calculations show that the supply of hydrogen in the sun will last for about 5 billion years.

5.
The Invisible Universe—The Energy of the Atom

The history of modern atomic theory; the atom, including a description of its nucleus and of the motions of particles within it; nuclear fission and fusion; and recent investigations of the atom

On July 16, 1945, the atom exploded with a cataclysmic force a thousand times more violent than the most powerful chemical explosive then known. This was "fission," the splitting of heavy atoms. Soon afterward came "fusion," the welding of the lightest atoms, with the resulting thermonuclear force of the H-bomb, a million times more powerful than the most powerful chemical explosive. When Albert Einstein realized the destructive use to which his equation $e = mc^2$ (energy equals mass multiplied by the square of the speed of light) could be put, he said, "I wish I had become a blacksmith."

To understand why in 1945 the atomic bomb became technologically possible, one must go back even before 1905 when Einstein developed his equation. Late in the nineteenth century the German physicist Wilhelm Roentgen had discovered X rays; the British physicist J. J. Thomson had discovered the electron; Henri Becquerel in France had prepared salts of uranium, an odd element and mineral nuisance, and had discovered radioactivity—rays spontaneously generated by the element itself; and Polish-born chemist Marie Curie had isolated the radioactive element radium. Finally, the New Zealand-born physicist Ernest Rutherford had already asked the multimillion-dollar question, "How can atoms (supposed to be ultimate particles of matter) give off rays?" He had been associated with Thomson in the researches that had led to the discovery of the electron and had gone to Canada at the age of twenty-seven to become professor of physics at McGill University, where, according to Sir Arthur Eddington, he initiated the greatest change in our ideas of matter since Democritus, four hundred years before Christ.

Rutherford established the existence of two emanations distinct from the electrons on which he had worked with Thomson. These were the alpha rays and the beta rays (the

"A" and "B" of the Greek alphabet). Alpha rays are now known to be the nuclei of helium, and beta rays are streams of electrons and positrons released from any source. Working on thorium, Rutherford established the concept of "half-life." In his own words (1906): "In the first 54 seconds, the activity is reduced to half value. In twice that time . . . the activity is reduced to one-quarter value, and so on . . . ."

In the intervening years Rutherford's description of atom decay has not been found wanting. One of his collaborators at McGill was Frederick Soddy, who discovered isotopes, species of elements that behave chemically like the dominant partner but have different physical attributes. One of Rutherford's students was the German Otto Hahn, who was, with his discovery of uranium fission thirty years later, to confound his teacher's positive statement, "The atom will always be a sink of energy and never a reservoir."

The trail of discovery of the atom followed Rutherford to Manchester, England, where another German student, Hans Geiger (whose name is now associated with the counter, or detector, of radiations), asked his teacher, "Don't you think that [Ernest] Marsden ought to begin a small research?" Rutherford said, "Why not let him see if alpha particles can be scattered [deflected] through a large angle?" This was something Rutherford should not have done to a young compatriot because he knew that the alpha particle was a fast, massive particle unlikely to be deflected. Marsden nonetheless fired a thin beam of alpha particles at metal foil. Three days later Geiger said to Rutherford, "We have been able to get some alpha particles coming backwards." Rutherford's later description of his reaction was typical. "It was quite the most incredible event that has ever happened to me in my life. It was almost as incredible as if you had fired a 15-inch shell at a piece of tissue paper and it had come back and hit you!"

This was one of the most significant scientific discoveries ever made. It showed that within the atom there was a prodigious force capable of deflecting an alpha particle traveling at 10,000 miles (16,000 kilometers) a second. But the recoil was infrequent, which meant that an immensely greater proportion of alpha particles was passing through the atom unimpeded. Following this experiment, Rutherford announced that he "knew what an atom looked like" and that he had an explanation for the deflection of the alpha particle. The atom, he had decided, must consist of a tiny, electrically charged, central particle in which practically all of the mass is concentrated, surrounded by a sphere of electrification, very thinly spread, of the opposite charge. (Later studies have revealed that the diameter of the nucleus is only about a hundred-thousandth of the whole atom. The remaining space is almost empty, and in this space the electrons move.)

Traffic Laws in the Atom

With Rutherford's experimental evidence and with help from the quantum theory, Niels Bohr established the fundamental laws governing the motion of electrons around the nucleus. The simplest situation is the hydrogen atom, which has only a single positive charge on its nucleus and a single negatively charged electron. The hydrogen atom's electron can move in one of a number of possible orbits, but, under the quantum rule, each orbit is specific or "permitted." The most stable of the allowed orbits is that of the lowest energy. In this track the electron keeps within one angstrom unit (one one-hundred-millionth of a centimeter) of the nucleus. There, the angular momentum (the amount of rotation of the particle in its orbit) is zero. At the next higher level of energy the electron is allowed any one of four possible orbits. At a still higher energy level the electron may have more than four allowed orbits.

Besides revolving around the nucleus the electron, like the Earth, also spins on its own axis. The electron's spin can take one of two directions, left or right. This doubles the number of permitted motions, or kinds of orbit, since in any given orbit it may spin in one direction or another.

In an atom heavier than hydrogen, that is, with more than one electron, the electrons' motions are much more complicated; just as the Earth's orbit round the sun is distorted from a perfect ellipse by the gravitational effects of other planets, so the motion of each electron around the nucleus of a heavy atom is influenced by the presence of other electrons. The Pauli exclusion principle, however, insists that two electrons with the same magnitude, direction of orbital angular momentum, and spins in the same direction can never exist in the same orbit. The "traffic laws" in the atom are well defined!

The Bohr-Rutherford theory of the atom in 1913 provided a "kindergarten model" of the microcosmos. It was simple: the atom was a nuclear "sun" with a planetary system of electrons. It disposed of the idea of atoms as "indivisible"; it helped chemists to explain how atoms linked up in molecules;

and, if the quantum theory gave meaning to the model, the model also gave meaning to the quantum theory.

For example, the single electron of hydrogen is lonely and restless. Hydrogen readily loses its electrons and becomes a positive ion (a "naked" nucleus) or proton. This is highly reactive and enters easily into chemical combination with other elements. In contrast, the helium atom is extremely stable. Its nucleus (Rutherford's favorite, the alpha particle) has two positive charges, which keep the two orbital electrons in tight conjunction with it. As a result helium has not yet been observed to react or combine chemically. The same is generally true of the other so-called noble gases—neon, argon, krypton, and xenon. Each is sustained by a closely knit system of electrons and does not readily part with any of them (ionize) or ordinarily associate with other elements. Indeed, the stability of these five gases is so exceptional that their atomic numbers, 2, 10, 18, 36, and 54 are known as the "magic numbers" of the Periodic Table. It came as a great surprise to the scientific community when successful syntheses of fluorides and other compounds of xenon and krypton were announced in the 1960s.

The picture of the atom's electronic structure enabled chemists to group elements in "families" and to predict their chemical behavior. But physicists at the time thought this to be superficial or "juggling with numbers." The Nobel Prize-winning British physicist Lord Rayleigh said, "I have looked at it, but saw that it was no use to me." Indeed, knowledge of the electronic structure of the atom gave little information or clues to the structure of the nucleus. The chemical nature of an atom is determined by the number of its electrons and protons, which must be equal in order to provide the balance of positive and negative charges. But that does not help to explain how the nucleus itself holds together. The forces that bind the nucleus together are millions of times greater than those that bind the electrons to the nucleus.

Probing the Nucleus

The idea of various nuclei consisting of combinations of protons was not satisfactory. It was all right for hydrogen—one proton and one electron, opposite in charge. The nucleus represents about 99.95% of the mass of an atom, however, which does not correspond to the mass of the protons necessary to give the opposite charge to a similar number of electrons; the mass for such a calculation is excessive.

Rutherford, in 1920, made some inspired predictions in his Bakerian Lecture to the Royal Society of London. He stated that it seemed likely that a nucleus could exist having a mass of two units and a charge of one unit, which would mean that it would behave chemically like hydrogen. In the United States, eleven years later, Harold Urey, Ferdinand Brickwedde, and George Murphy discovered just such an atom—deuterium, the isotope of hydrogen, or "heavy hydrogen." Rutherford also assumed the existence of a particle with a mass of three units and a charge of two units, which materialized in later experiments as a lighter isotope of helium. Most remarkable, however, was his anticipation of that portentous particle, the neutron. It would have no electric charge. "It should be able to move freely through matter," he told his colleagues. "Its presence would probably be difficult to detect by spectroscope, and it may be impossible to contain it in a sealed vessel. On the other hand, it should enter readily the structure of atoms, and may either unite with the nucleus or be disintegrated by its intense field, resulting possibly in the escape of a charged hydrogen atom or an electron or both."

Twelve years later (1932) James Chadwick, at the Cavendish Laboratory at Cambridge, established the existence of such a particle, which was given the name "neutron." This ghost particle, which, having no electric charge, could move anywhere and enter—as Rutherford had foreseen—into the nucleus of the atom, became the "trigger" of the atomic bomb.

In those hectic days of the early 1930s, in addition to the discovery of the deuteron (the nucleus of the deuterium atom) and the neutron, the U.S. physicist Carl Anderson discovered the positron, the positive electron, which another of Rutherford's young researchers, Paul Dirac, had theoretically postulated; Nobelists Sir John Cockcroft and Ernest Walton had used their high-voltage accelerator (compounded of packing cases, cookie tins, and glass tubes and sealed with plasticine) to split the atom; Frédéric and Irène Joliot-Curie in Paris had produced artificial radioactivity by bombarding boron and turning it into a form of nitrogen that gave off rays (they also exposed uranium to a bombardment of neutrons with curious results that were to assume a more significant meaning later); Enrico Fermi in Italy had named the neutrino; and Hideki Yukawa in Japan had described the meson.

The neutrino had to exist, once the character of the neutron had been established. Inside the nucleus a neutron can live indefinitely, but when the particle is observed outside it proves unstable. In an average time of about eighteen minutes it spontaneously ejects a beta particle (a nuclear electron) and turns into a proton. The proton and the electron are about 1.5 electron masses lighter than the neutron, and so this amount of mass, equivalent to 780,000 electron volts of energy, appears to be lost. The Austrian-born physicist Wolfgang Pauli suggested that the discrepancy might be accounted for by another particle, almost undetectable. Fermi pursued this surmise and in 1934 constructed a complete theory of beta decay. Its fundamental process is that a neutron continuously loses and regains an electron and a neutrino by emission and absorption.

Yukawa set out to describe the "glue" that held together the protons and neutrons in the nucleus. He proposed that jointly they emitted and absorbed a particle called a "meson." Its force would extend only over very short range, and it would have a finite mass.

Three years later a meson did materialize in the cosmic rays from outer space, detected on Earth. It seemed to have just the properties that Yukawa had specified. It had a mass about two hundred times that of the electron and was found to have positive and negative forms, but it was not what the scientists were looking for because it did not react strongly with the other particles, the protons and neutrons, and therefore could not transmit nuclear forces. Much later, in 1947, Yukawa's specifications were met by another type of meson (trapped in the emulsion of photographic plates sent up by balloon to high altitudes). C. F. Powell, a Briton, G. P. S. Occhialini, an Italian, and C. M. G. Lattes, a Brazilian, discovered that this new particle did interact strongly and had a mass of 273 electrons. This was the pi meson or "pion."

Rutherford died in 1937, having imprinted his personality on the whole of nuclear research. If at that moment a scientifically inquisitive Alice had sipped her "Drink me" and had slipped into the Wonderland of the nucleus, she would have found it tidily furnished with the electron, the proton, the positron (the positive electron), the neutron, the neutrino, and the meson. In varying numbers and combinations, those "elementary particles" could account for the structure of the (then) known ninety-two elements. Only a few of those elements were unstable, and their instability could be satisfactorily explained by the eccentric behavior of the particles.

The Atom Explodes

In 1919 Rutherford had shown that by using alpha particles spontaneously released from radium, he could split the nitrogen atom, expel a hydrogen nucleus (or proton), and convert nitrogen into oxygen. He had proved what the medieval alchemists had believed—at the risk of their being broken on the rack or burned at the stake—that substances could be transmuted; he had founded modern alchemy. In 1932 Cockcroft and Walton had devised the high-voltage accelerator, which applied about 750,000 volts to a proton and then used it as a projectile to hit a lithium target. The proton split the lithium atom, thereby releasing energy from it of 16 million electron volts. This seemed a substantial dividend, but Rutherford pooh-poohed it as a potential source of useful energy. He pointed out that only one proton projectile in 10 million hits the target. He said, "It's like trying to shoot a gnat on a dark night in the Albert Hall and using 10 million rounds of ammunition on the off-chance of getting it." Meanwhile, Ernest Lawrence in the United States had improved on the atom smasher by devising the cyclotron, in which particles magnetically directed into an endless circle could generate higher and higher velocities and, therefore, higher and higher energies. But it was still "hitting a gnat on a dark night."

All that was changed within two years of Rutherford's death. It has been mentioned that in 1905 the young German chemist Otto Hahn had chosen to be Rutherford's student at McGill University; it has also been mentioned that the Joliot-Curies had exposed uranium to a bombardment of neutrons with curious results. Just how curious emerged from Hahn's experiments (with his colleague Fritz Strassmann at the Kaiser Wilhelm Institute in Berlin) when he irradiated uranium with neutrons and found that there was transmutation into two elements, showing that the uranium had split as a result of the intervention of the neutron. The Austrian-born physicist Lise Meitner and her nephew, Otto Frisch, rightly interpreted this as "fission" (a term borrowed from biology, where it refers to cell division). There was a further significant fact that the fission released a neutron that could split other uranium atoms. This could produce a chain reaction.

Chain reaction simply means that one event will produce another and another and another. The end result is familiar to the whole world. If a chain reaction can be sustained—if neutrons from one uranium atom can be "captured" by the nucleus of another uranium atom—they will produce another split and the release of more neutrons. There also will be a release of surplus energy. If the process can be sustained within a given amount of atoms (critical mass) and in an instant of time, the result will be an explosion.

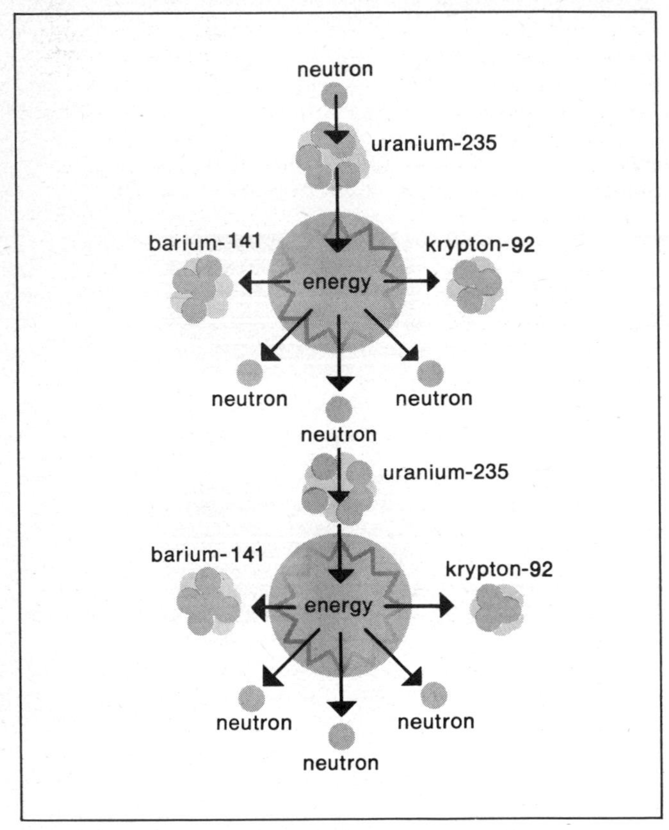

A nuclear chain reaction begins when a neutron (top) splits an atom of uranium-235, releasing energy and more neutrons. These neutrons then collide with other uranium atoms, continuing the reaction until the desired amount of energy is obtained.

It was recognized that the operative particle in the reaction, the particle that was spontaneously releasing neutrons, was uranium-235, which occurs in natural uranium in proportions of 1:140 atoms of uranium-238. If enough ^{235}U could be separated, it would produce an explosion, but, as Fermi demonstrated in the atomic pile at the University of Chicago, it was also possible to arrange the process so that the fast neutrons of ^{235}U could be slowed down enough to be captured by atoms of ^{238}U. This produced a man-made element, plutonium, which was unstable and in the proper quantity would also produce an explosion.

It thus became a question of arranging atoms in a "lattice" so that neutrons would be captured. In a bomb the instantaneous release of energy would be catastrophic, but under control it could produce peaceful atomic energy.

The next step was the hydrogen bomb, which depends on the thermonuclear process, which is the means by which the sun generates and releases energy. The process depends not on splitting atoms but on fusing them. The nucleus of the hydrogen atom consists of one particle. The nucleus of the helium atom consists of four particles. If the heat is sufficient (like the 20,000,000° C in the heart of the sun) four particles of hydrogen can be made to fuse (to become helium) with a surplus energy a million times greater than chemical energy and a thousand times greater than fission energy. Such temperatures can be produced on the Earth in the instant of the explosion of a fission bomb. The trick, therefore, was to use the fission bomb as the "percussion cap" and to surround it with susceptible material. As we know, this instant "pressure cooker" worked to produce the hydrogen bomb. Another dragon had been released, fiercer than that of the fission process; attempts were soon made to tame it.

The peaceful uses of fusion energy were not as simply arrived at as were the peaceful uses of fission energy. As president of the 1955 United Nations Conference on the Peaceful Uses of Atomic Energy, Homi Bhabha said that if it could be tamed thermonuclear energy would provide as much industrial power "as there is deuterium in the seven seas." It was then discovered that little was actually known about the behavior of the particles, and to remedy this yet another new science came into being, plasma physics.

This is an interesting example of how words become corrupted in scientific usage. *Plasma*, to a Greek scholar, means "mold" or "matrix." When it was first adopted by the biologists in such terms as *protoplasm*, it respected its origins. When, however, it became familiar to blood donors, it no longer meant "matrix," but what went into the matrix; it meant the blood fluid without the corpuscles. Finally, when the physicists adopted it, *plasma* was neither a matrix nor the fluid without the corpuscles; it was the corpuscles. *Plasma* meant the particles separated from each other.

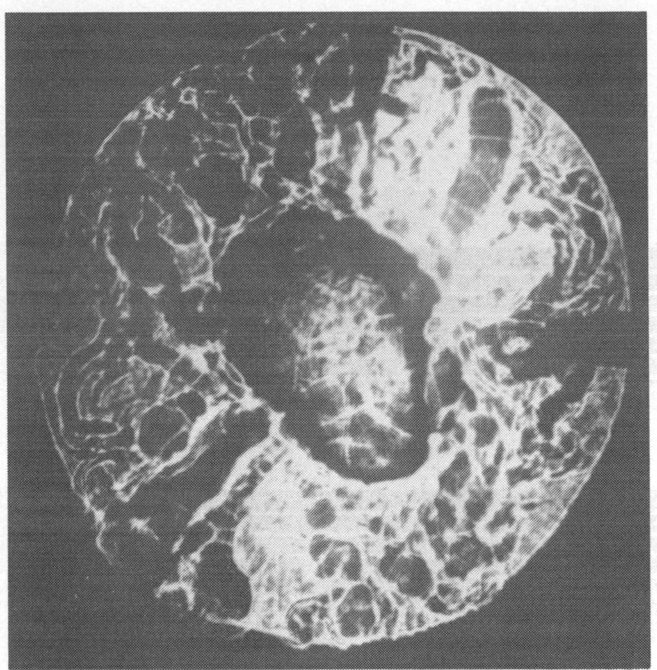

A plasma is a mixture of negatively charged electrons and positively charged nuclei. Matter in the plasma state can undergo thermonuclear reactions that produce vast amounts of energy.

Plasma was used to describe the "fourth state of matter"— not solid, not liquid, and not gas, but something else. If scientists had called it "corpuscular flame," they might have raised the phantom of phlogiston (once believed to be the hypothetical principle of fire of which every combustible sustance was in part composed). But, just as it was possible to extract free electrons (electric currents) directly from flame, so in plasma physics it was sought to generate the temperatures necessary for thermonuclear fusion by the direction of particles. This management of corpuscles was already familiar to lighting engineers in the form of strip lighting (such as neon signs) in which beams of electrons are passed through gases in vacuum tubes to produce luminescence. In thermonuclear applications, however, the ambitions far exceeded those of lighting engineers. Physicists were seeking temperatures far, far greater than the temperature of the heart of the sun (because, as has been already stated, the sun is a slow oven with billions of years in which to process the fusion). Terrestrial materials cannot stand such temperatures—neither metals, nor glass, nor ceramics. If, therefore, the beam of particles were to impinge on the walls of any such system, the materials would disintegrate. In plasma engineering, therefore, magnetic fields, to keep the beam away from the walls, are the invisible "wadding."

The Safecrackers and the Locksmiths

The safecrackers forced the lock of the atom before the locksmiths knew how it worked. This seems a disrespectful way of describing the greatest material achievement of man since our ancestors mastered fire. Moreover, the safecrackers included, directly or indirectly, almost all of the nuclear locksmiths of the Western Allies, including the founding fathers, Einstein and Bohr, who gave their insight and their influence. With the greatest muster of scientific brainpower ever invoked and $2.5 billion worth of technology, the Manhattan Project released the energy of the atom.

In 1940 Rudolf Peirls and Otto Frisch had sent a report to a British committee that was examining the possibilities of an atom bomb. They were able to say that a chain reaction could be sustained and a superbomb produced. Hahn and Strassmann had provided the clue. Meitner and Frisch had "read the message" in terms of fission. Finally, Bohr and the U.S. physicist John Wheeler had proposed their "liquid-drop" model of the nucleus.

The "liquid drop" was an ingenious evasion of all the unanswered questions. It was recognized that the forces that bound the nucleus were millions of times greater than those that bound the electrons to the nucleus. But no one knew how those forces were created, and even if that had been entirely known, scientists would still have been confronted with the prodigious difficulty of calculating the results of those forces upon a large number of protons and neutrons that interacted with one another strongly and at extremely short distances. The Bohr-Wheeler version simply treated the nucleus as though it were a drop. Just as in a water molecule the chemical identities of hydrogen and oxygen are merged, so in the nuclear "drop" the protons and neutrons lost their identities. And the "drop" lent itself nicely to the concept of fission: the drop could elongate like a raindrop on a windowpane, with a waistline that diminished to a snapping point.

Needless to say, the mathematics and physics of the Bohr-Wheeler "drop" were not as simple as the image. How the invading particle, the neutron, upset the homogeneous proton-neutron "drop" and caused it to divide and how in that division spare neutrons were released to repeat the process were matters of laborious and ingenious calculation and experiment. Nevertheless, this was high-level know-how that was ultimately translated into technological know-how to produce the nuclear fission bomb.

With deference, however, one can repeat that these scientists had cracked a lock of which they did not know the combination. For example, they were then conceiving of the nucleus as consisting of protons, neutrons, beta particles, gamma rays ("hard" X rays), and Yukawa's "glue," the gripping mesons.

J. Robert Oppenheimer, as head of the laboratory at Los Alamos, N. Mex., assembled his colleagues' knowledge, brought key men together in the actual operations, and compounded knowledge, experience, and material into the bomb that exploded at Alamogordo, N. Mex., on that morning in 1945 that changed world history. In 1958 when, with Bohr, he officiated at the opening of the Institute of Nuclear Science at the Weizmann Institute of Science in Israel, Oppenheimer said, "If you had asked me ten years ago what the structure of the nucleus looked like, I might have told you. Ask me ten years from now, and I may be able to tell you. At the moment, I cannot tell you."

Between July 1945 and 1958 the tremendous impetus to nuclear research that the nuclear bomb had produced had increased the known components of the nucleus to thirty. Scientists still referred to them as "elementary particles," and though they were detectable only individually they could nevertheless be conceived as forming an orderly pattern and a consistent relationship. Within the next five years, however, seventy other subatomic objects had been discovered and, with this embarrassment of riches, the scientists began to paraphrase George Orwell's *Animal Farm* statement: "All animals are equal, but some animals are more equal than others." All nuclear particles are elementary, but some must be more elementary than others.

In 1952 the Austrian physicist Erwin Schrödinger was already saying: "Fifty years ago science seemed on the road to a clear-cut answer to the ancient question, 'What is matter?' It looked as though matter would be reduced at last to its ultimate building-blocks—to certain submicroscopic but nevertheless tangible and measurable particles. But it proved to be less simple than that. Today a physicist can no longer distinguish significantly between matter and something else. We no longer contrast matter with forces or fields as different entities; we know that those concepts must be merged."

Because of their success with the atom bomb, nuclear physicists after World War II were at the top of the hierarchy of science. What they asked for they got. And they asked for plenty. A long way from the sugar crates and plasticine of Cockcroft and Walton, the postwar generation of linear accelerators and the various "—trons" (in the succession to Lawrence's cyclotron) were gargantuan and ran into billions of dollars of expenditure. These were not (as they were sometimes called) "atom smashers"; rather they were "atom builders" that used energy to locate mass. The highest energies were attained by accelerating protons.

In 1972, outstripping the Soviet Union's Serpukhov, a powerful particle factory near Moscow, the Fermi National Accelerator Laboratory, or the Fermilab, began operations at Batavia, Ill., near Chicago. It consisted of more than a thousand powerful electron magnets situated around a circular vacuum tube nearly 4 miles (about 6 kilometers) in circumference. Protons circulated 200,000 times around this track, traveling a distance farther than to the moon and back. At intervals, those protons were channeled into laboratories as big as factories to be used in dozens of different ways. In 1976 a comparable machine was built at CERN, the European nuclear research center at Geneva, Switzerland.

In these machines the nuclear physicists were playing with the four (conventional) cosmic forces: (1) gravity, the builder of the stars, pilot of the planets, and manager of matter—acting on all particles without discrimination; (2) electromagnetic force, the author of light, lightning, and life—working between particles possessing electric charge; (3) strong nuclear force, the binder of the atomic nucleus, and fire maker of the sun and stars—operating on particles at close quarters;

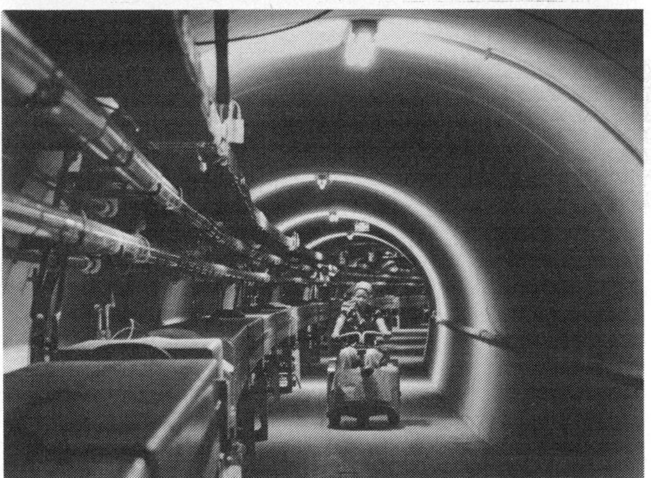

The main accelerator or "Main Ring" at the Fermi National Accelerator Laboratory (top) measures nearly 4 miles in circumference and 1.24 miles in diameter. Inside the "Main Ring" (bottom) 1,014 bending and focusing magnets guide proton beams through a vacuum tube along the circular course.

and (4) weak force, weak in strength but profound in influence—affecting the inherent qualities of matter, creating diversity, and appearing in a variety of radioactive decay processes. Gravity and electromagnetic force are unlimited in range, with electricity stronger than gravity. The strong nuclear force, however, is a hundred times stronger than electromagnetic force. The weak force operates only when particles are less than a million-millionth of a centimeter apart.

Physicists after 1970 were convinced of the existence of two more cosmic forces. One was a variant of the weak force, which turned out to have profound implications for the behavior of stars. The other was the strongest force of all, which has been called "color" although it has nothing to do with the optical wavelengths that give us color in the sense in which we ordinarily use the word. It operates between quarks.

The word *quark* illustrates the bewildering state that nuclear physics had reached by the 1960s. Instead of the nice tidy nucleus of electrons, protons, and neutrons that Rutherford had left (and which the bomb makers had exploded), there were countless manifestations that physicists tried to measure and label. Conventionally, they would have given them Greek symbols or names, but they had run out of nomenclature.

The Growing Family of Quarks

Murray Gell-Mann of the California Institute of Technology, one of the theorists trying to bring order out of the subnuclear chaos, had given the name "strangeness" to the quality of matter that enabled some particles to live longer than expected. He began grouping "strange" and "nonstrange" particles with mathematical families. The sorting out indicated that the proton was not a truly basic particle but was made of something else that could be bound more or less tightly together. Gell-Mann declared that all heavy particles were made out of three kinds of something. To find a name he went to the Irish novelist James Joyce and his phrase "Three quarks for Muster Mark." Gell-Mann was asking for trouble because *quark* in German means "cream cheese" or "nonsense." But the name stuck, and the experimentalists with their gargantuan machines started hunting for quarks.

Gell-Mann called the three types of quark "up," "down," and "sideways." The directions had no physical meaning. The quarks carried electric charges. The up-quark was said to have a charge equal to two-thirds of the charges on the proton and the down-quark and the sideways-force each had a negative charge of one-third.

Two up-quarks and one down-quark made a proton. There could also be antiquarks. An antiproton, for example, consisted of two anti-up-quarks and one anti-down-quark. A neutron turned out to be an up-quark and two down-quarks. Transmuting a proton into a neutron involves changing an up-quark into a down-quark. Mesons were also made of quarks, one quark in conjunction with an antiquark. This would be consistent with a meson's behavior as a short-lived force-carrier.

For some time physicists could get along with three quarks. When large particles were broken apart, the three-quark theory predicted the particles that would be produced, but in the late 1960s experiments revealed four types of reactions that the three quarks could not explain. For instance, according to the three-quark theory, a K meson should decay into a pi meson plus an electron and a positron, but the pattern was not there in the smashed atoms. Besides, with three quarks there was no way of unifying the three forces of nature that govern subatomic particle interaction. Scientists believed that a fourth quark might do it. It would need to be heavier and longer lived than any of the other three.

In November 1974 researchers at Stanford University's Linear Accelerator Center in California found their instruments becoming boisterous. The scribblings of the recording pen bounced off the paper, giving evidence of the heavy particle for which they were looking. The Stanford scientists called the particle "psi." By another technique the Brookhaven National Laboratory in New York had simultaneously discovered a similar particle and had called it "J." So, in fairness to both groups, the particle became known as "J/psi" or "gypsy." It is three times as massive as the proton and also has its anti-quark.

Moreover, gypsy was consistent with "charm." In the whimsical nomenclature that followed the exhaustion of the Greek alphabet, charm, according to Sheldon Lee Glashow, "is the magical device to avert evil." The "evil" was the continued failure to unite the weak force with the electric force. There were three quarks but four members of the electron family—the ordinary electron, the heavy electron, and two distinct types of neutrinos. Wouldn't it be "charming" if there were a match, symetrically four to four, between the two families of quarks and electrons?

The three forces that govern all of the known interactions of subatomic particles are the electromagnetic force, the strong nuclear force, and the weak force. (Gravity exerts no appreciable effect on particles.) The "charmed" existence of gypsy can help physicists unify these three forces into one theory. The practical implications, though a long way away, could be as immense as the discovery of electricity—the development of entirely new materials, new forms of energy, new access to space, and a new understanding of the universe.

But physicists cannot leave well enough alone. After charm we now have "truth" and "beauty." Haim Harari of the Weizmann Institute in Israel envisaged two additional pairs of particles, making a total of six quarks and six members of the electron family. The proposed additions to the electron family included a super-heavy electron (a heavy "lepton" and a new neutrino to go with it). Harari suggested that the quarks be labeled "top" and "bottom," but in the new poetry of physics they have become "truth" and "beauty." In July 1977 at Stony Brook, N.Y., Leon Lederman announced the discovery of a new heavy particle that, reverting to Greek, he called "upsilon" with 9,000 mass-energy units, and it beautifully paired off with a super-heavy electron found at Stanford.

A team of physicists led by Peter Kotzer of Western Washington State College, conducting an experiment in December 1978 in connection with the Fermilab, "saw" neutrinos for the first time. After detecting about 3,200 a day, Kotzer said that "When you're looking where no one has looked before, you see surprises—our surprise is that we got so many . . . ." He then speculated that yet another new particle and an unknown reaction were involved.

6.
Spaceship Earth—Its Magnetic and Gravitational Energy

The Earth and its atmosphere, including the Van Allen Belts and the auroras; the Earth's magnetism and its interior structure; and the theory of plate tectonics

"Spaceship Earth" is not a whimsical term; it is a useful description. The planet is an energized capsule, wrapped up in its atmosphere and invisibly encased by its own magnetic and gravitational forces. It is a vehicle (with several billion passengers on board) traveling at 18.5 miles (nearly 30 kilometers) per second, or 64,600 miles (more than 107,000 kilometers) per hour, in its 365.25-day orbit round the sun.

Although the gases of the atmosphere can be detected hundreds of miles above the Earth's surface, the major part of the gaseous matter is in a zone 5–11 miles (8–18 kilometers) in thickness; thus, the Earth's inhabitants exist at the bottom of an ocean of air. The part of the atmosphere in which we spend our working lives (except for jet pilots) is the troposphere. It is here that most clouds form and provide our weather. In the troposphere temperatures generally decrease with height, and there is a considerable amount of turbulent mixing. Water vapor that enters the atmosphere by evaporation from the surface and leaves it as rain, snow, hail, dew, or hoarfrost is trapped in this region.

The upper boundary of the troposphere is the tropopause, about 6–8 miles (10–13 kilometers) above the Earth's surface. Above it is the stratosphere. The latitudinal variation of temperature in the stratosphere is smaller than in the troposphere. At an elevation of 12 miles (20 kilometers) the January temperature increases from about −185° F (−120° C) at the equator to −67° F (−55° C) or higher around 55° N and then decreases to about −100° F (−75° C) at the North Pole. In July at the same height the temperature rises continuously from −75° F (−60° C) at the equator to −40° F (−40° C) at the North Pole. Above 18 miles (30 kilometers) the temperature increases again because of the absorption of ultraviolet radiation by ozone, and at about 30 miles (50 kilometers) the temperature reaches a maximum of 32° F (0° C). This is the level of the mesosphere, where the temperature again decreases. The air density at this level is only 1% of that at sea level. The characteristic stratospheric temperature distribution is due to the interaction of solar and terrestrial radiation.

Strong jet winds are a feature of the stratosphere. During January there is a strong westerly jet with a maximum of 100 knots centered at the tropopause at 40° latitude. Another westerly jet with a maximum of 80 knots operates at an altitude of 15 miles (25 kilometers) at 70° latitude. In the same season an eastern current of 40 knots prevails at 20° latitude at an altitude of 15 miles. In midsummer at mid-latitudes the westerly jet at the tropopause has a speed of 50 knots, and at higher altitudes there are easterly currents at all latitudes.

Above the mesosphere is the ionosphere, critically important for radio communications. This is an electrically conducting sphere entirely surrounding the Earth. It has been called "a mirror in the sky" because it reflects radio signals. The ionosphere consists of two principal regions of free electrons, produced by the sun's ultraviolet radiations. The lower, the E-region, normally occurs at heights of about 55–70 miles (90–120 kilometers), while the upper, the F-region, is usually found at 90–180 miles (150–300 kilometers). Rocket observations have confirmed that the ionosphere is highly ionized, that is, that it carries an electric charge. The characteristics of ionospheric layers change with the time of day, season, and geographical location. Variations in the E-layer and the lower parts of the F-layer are generally small, but abnormal ionization of the E-region can occur in the summer. Fluctuations in the upper levels of the F-region are often large, irregular, and most pronounced during magnetic storms.

The eleven-year sunspot cycle and other forms of solar disturbances affect the behavior of the ionosphere. That behavior is important because the ionosphere controls long-distance radio communications by dictating the upper and lower limits of usable wave frequencies. The upper limit is determined by its reflecting properties and the lower limit by absorption in the lowest part of the ionosphere, the D-region. During periods of solar disturbances the enhanced ultraviolet and X-ray emanations cause ionization of the D-region, resulting in an increase in absorption and the muffling of radio waves, which results in poor radio communications.

Structure of Earth's Atmosphere

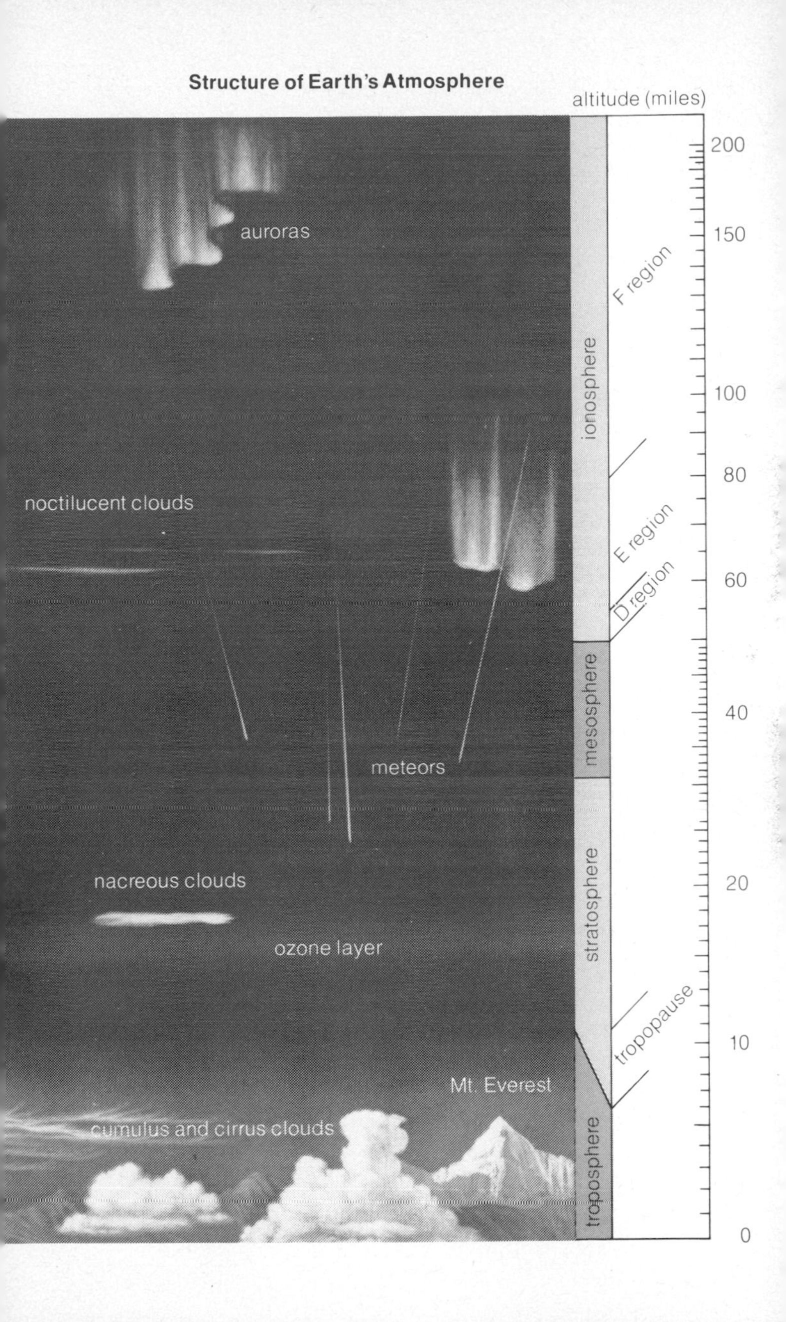

The Van Allen Belts and the Auroras

Before space science man was limited in his exploration of the Earth and its atmosphere up to an altitude of about 25 miles (40 kilometers). After World War II space rockets (based on the rocket weapons that the Germans had developed to bombard Great Britain from the continent of Europe) were used as probes. The warheads were replaced by scientific instruments that could measure the characteristics of the upper atmosphere and the solar and cosmic radiation. Such rockets had a brief lifetime, however, so the amount of information that they could relay back to Earth was limited.

In the biggest cooperative scientific investigation ever attempted, scientists of more than seventy countries, in greater or lesser degree, combined in the International Geophysical Year (IGY). About thirty thousand scientists and observers, operating several thousand scientific stations in a network stretching from pole to pole, were ultimately involved. The object was to use the year of the sunspot maximum, 1957–58, as the basis for a systematic study of the Earth and its environment. The "year" was extended to thirty months, to Dec. 31, 1959. Eleven fields of geophysics were involved: aurora and airglow; cosmic rays; geomagnetism; glaciology; gravity; ionospheric physics; longitude and latitude determinations; meteorology; oceanography; seismology; and solar activity.

The IGY committee proposed the use of artificial satellites to patrol space beyond the Earth's atmosphere, which limits observations. The Soviet Union and the United States both agreed to contribute satellites for this effort.

The U.S.S.R.'s *Sputnik I* was first into orbit on Oct. 4, 1957. *Explorer I* was launched by the United States on Jan. 31, 1958. Included in the equipment of *Explorer* was a Geiger counter installed by the U.S. physicist James Van Allen, which reported back surprising evidence of an intense belt of radiation surrounding the Earth. Subsequent experiments with other satellites and with *Pioneer* space probes provided confirmation. It has been found that there are an inner belt and an outer belt, zones of charged particles trapped in the magnetic field of the Earth. The zones are most intense over the equator and weakest over the poles. The particles are mainly protons and electrons. Many of the inner-belt protons are so intense that they can traverse several inches of lead. The inner ring is doughnut-shaped and extends from about 500 to 1,000 miles (800 to 1,600 kilometers) above Earth; the outer extends from about 900 to 1,800 miles (1,500 to 3,000 kilometers).

There is clearly a connection between the auroras and the Van Allen Belts. The Northern (borealis) and Southern (australis) Lights are the luminous draperies that appear in the night sky in high latitudes. Their lower fringes are about 60 miles (100 kilometers) above the Earth's surface, and the arcs and rays extend upward for another 100 miles (160 kilometers). During the International Geophysical Year, an array of 120 all-sky cameras (90 in the Northern Hemisphere and 30

Charged particles from the sun (the solar wind) contain the magnetosphere and give it a definite boundary. Some of these particles are accelerated within the magnetosphere and guided to the polar regions, where they interact with atmospheric gases to cause auroras.

in the Southern) showed that the auroral arcs extend along at least 3,000 miles (4,800 kilometers) from east to west.

The auroral light is caused by charged particles entering the atmosphere and spiraling round and descending along the lines of magnetic force. The particles ionize atmospheric atoms and molecules by colliding with them, the impact occasionally knocking off an electron. The particles also excite atmospheric gases. The yellow-green light that frequently dominates the visual color and also the red radiation seen at higher levels are produced by atomic oxygen. Rockets flown into active auroras have confirmed the observations of the behavior of particles.

Atom bombs can produce man-made auroras. Three nuclear explosions (the Argus experiments) in 1958 produced artificial belts that decayed in about a week, but the Starfish explosion of 1962 had effects that persisted much longer and extended outward to about 12,000 miles (20,000 kilometers).

The Magnetosphere

Through observations and positive tests, we now have a pretty convincing picture of how Spaceship Earth performs. It is not just a ball hurtling through a void. It has its envelope of atmosphere; its electric sheath, the ionosphere; and a magnetic "hull," within which the Van Allen Belts serve as bulkheads. At about 40,000 miles (63,000 kilometers), ten times the Earth's radius, the Earth's magnetic field stops quite abruptly. Unlike a bar magnet in a vacuum the field does not decrease indefinitely with increasing distance. The Earth is not surrounded by a vacuum but by the outer atmosphere of the sun. The sun's corona, with its extremely high temperature, is unstable and continuously expanding. The result is a solar wind. This is an ionized gas blowing at an average velocity of 300 miles (500 kilometers) per second. Blowing at all times, the wind is composed mainly of protons with energies of about 1,000 electron volts and electrons with energies of about 10 electron volts, flowing radially from the sun. The existence of this wind was first recognized by observations of the tails of comets and was later measured directly by man-made satellites.

The wind pushes the magnetic field in and is deflected by it. A boundary is thus formed when the deformation of the magnetic field just balances the pressure of the solar wind. The Earth's magnetic field is, therefore, confined within a region called the magnetosphere. The geomagnetic field lines

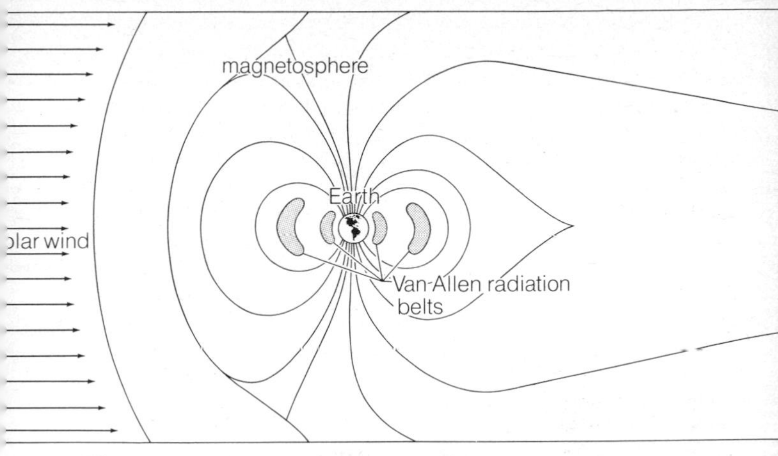

The magnetosphere, a volume surrounding the Earth, contains a huge magnetic field that differs in intensity from one part to another. Part of the field is pushed toward the Earth by an enormous solar wind that contains charged particles. Another part is pushed away from the Earth to form a "tail." The Van Allen radiation belts—two regions of intense radioactivity—are doughnut shaped and encircle the Earth.

flow behind it like a magnetic wake, similar to a comet's tail.

The Earth, with its magnetosphere, is a supersonic spaceship traveling 64,600 miles (more than 107,000 kilometers) an hour. Just as a supersonic aircraft is preceded by a shock wave, so the magnetosphere produces a bow shock wave in the solar wind. This wave has been measured by magnetometers and solar-wind detectors. And just as a supersonic aircraft causes the air ahead of it to heat up and flow turbulently around it, the transition zone between the bow shock wave and the magnetosphere is found to contain heated protons careering around in confusion, like air turbulence.

View from Space

The men who landed on the moon and those who circled the Earth in man-made satellites have reminded us how relatively small our planet is. In the celestial domain the stars (of which the sun is one) are thought to be more numerous than the grains of sand on all the beaches of the world, and billions of them have planets.

What the eyes, human or electronic, in orbit see is the

family estate of the human race. The optic mosaic, like the TV camera mosaic, registers a picture of a slightly distorted sphere, somewhat similar to a classroom globe in sculptured relief—but not a very emphatic relief because from an altitude of, say, 125 miles (200 kilometers) the great mountain ranges are just wrinkles and the Grand Canyon is a scarcely visible crease. The eyes in orbit register the superficial features of a globe that is seven-tenths water and three-tenths land. Of that three-tenths (the continents and islands) about two-fifths are covered by hot and cold deserts and mountains, one-third by forests, one-fifth by permanent pastures, and about one-tenth by cropped lands. Thus the rangelands and cultivated lands on which the inhabitants of the Earth depend for their sustenance are a minor part of the total estate.

The natural atmosphere of the Earth (as distinct from the man-made atmosphere created by imperfectly combusted fossil fuels and chimney effluents) consists of nitrogen, oxygen, argon, carbon dioxide, neon, helium, krypton, xenon, hydrogen, methane, and nitrous oxide. This gaseous envelope is influenced by the Earth's living beings because they exude vaporous material and, in the process of living and dying, convert and liberate chemicals from the soil. Soil may be basically a product of rocks, being eroded by the rains and winds, themselves the violent forces of the sun-driven atmosphere—but the soil is also "living." A lump of soil no bigger than a football contains more organisms than there are people on Earth. Humus, the decay product of living materials, is an essential component of the topsoil on which all growth depends. Organisms in the topsoil break down the living materials and release the chemicals that plants need as nutrients and also make the "crumbs" of soil that hold moisture and help the roots of plants to "breathe." From this topsoil grow the plants that provide the food for animals and humans and at the same time provide the vegetation texture that protects the soil from erosion.

The Crust and the Core

All these are subject to energy radiations generated by the sun, which by remote control dictates the environmental conditions of the Earth. The sun has comparatively little effect, however, on the solid part of the Earth, influencing it chiefly by the surface forces that create erosion and carve and grind the rocks and by the deposition of the carboniferous layers of primeval living materials, which have stored the sun's energy as hydrocarbons. Otherwise, the crust of the Earth and its core are self-contained. If one accepts the tenet that the Earth was formed out of nebular gases and dust that were condensing, fragmenting, and again accreting, and rejects the theory that it was extracted from the sun, then it had a separate creation.

Divisions of the Earth

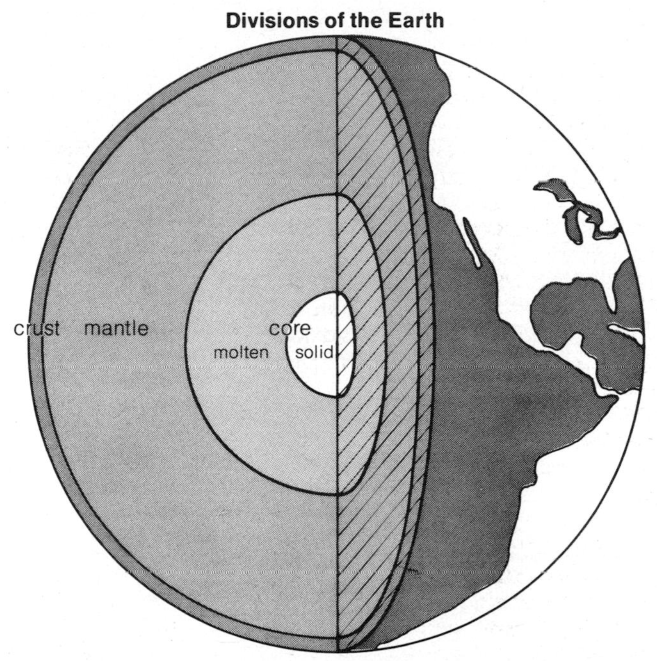

The Earth has an outer crust with a mean thickness of about 27 miles (43 kilometers), an eggshell compared with the dimensions of its envelope or the size of its interior. The crust is extremely thin below the oceans, usually less than 10 miles. Beneath this is a mantle that surrounds the central core. The Earth is like a golf ball—the dimpled casing (the crust), the different layers of elastic windings (the mantle), and the fluid sac in the center (the core). The radius of the Earth is about 3,900 miles (6,400 kilometers). One of the early discoveries of seismology, the study of earthquakes, was the discovery of the molten core and its dimensions.

There are two types of wave that travel from the site of an earthquake downward into the Earth. They are the primary

(P) waves, in which the particles of the Earth's material move longitudinally as the shock waves pass by, and the secondary (S) waves, in which the particles move sideways. In solids, P waves move 50% faster than S waves. P waves take 20 minutes and 12 seconds to go through the Earth to the other side. S waves do not transmit through fluids to any significant extent. Therefore, those parts of the interior found to transmit both P waves and S waves are solid. Failure to detect S waves is evidence that the part of the Earth involved is in a fluid state.

In this way seismologists have located the Earth's fluid core 1,785 miles (2,880 kilometers) below the surface. In 1936 this was further defined by the seismological discovery of an inner core with a 795-mile (1,280-kilometer) radius. Thus, apart from the crust, there are three major zones in the interior of the Earth.

The lower boundary of the crust is called the Mohorovicic Discontinuity after its Yugoslav discoverer. Below this is the mantle, which extends to the boundary of the outer core. The outer core is fluid (in a molten state), and the inner core is dense, most likely consisting of iron that has solidified under the velocity of shock waves, which vary with depth. There is a progressive stiffening of the material according to increasing pressure. In the mantle the pressure rises rather regularly at a rate of about 470 atmospheres per kilometer. The pressure at the boundary of the core is 1,370,000 atmospheres, and at the center it is 3,700,000 atmospheres.

The aggregation of the planet from nebular dust and fragments has as a theory replaced earlier speculations about its origin as a ball of molten lava erupted from the sun. The materials from which the planet was formed must have been relatively cold. The original Earth throughout billions of years was (on present assumptions) nonmolten. It could have come near to melting at a later stage as a result of the heat of the radioactivity generated by the fissionable atoms inherent in its own structure. At some time the Earth must have been soft enough to allow the separation of the various layers that seismologists can distinguish.

Radioactive substances generating heat tend to associate with the lighter elements. On the Earth such elements are mainly present in the outer layers, notably the crust. This would account for the solid center, the homogeneity of the lower outer core, and the nonhomogeneity of the upper parts of the outer core. It would be like cooking a Baked Alaska—

the ice cream remains solid in the center, while surrounded by a filler, with crisp meringue on top.

Chemical sampling of rocks can provide direct insight into the nature of the Earth's crust. The findings can be compared with the chemical analyses of the meteorites that puncture the "dust sheet" of the atmosphere and that may be the fragments of a planet, or embryonic primordial planets that failed to mature. Another comparison can be made with the chemical elements of the sun and other stars as revealed by spectroscopy. The results from all of the examined sources—Earth, meteors, sun, and stars—are reasonably consistent and support the contention that the gross chemical composition of universal matter is uniform. "Cosmic abundances" of elements have been estimated, using silicon as the reference element because rocks consist largely of compounds of silicon.

Harold Urey, a U.S. Nobel Prize-winning chemist, and others worked out the relative abundance of elemental atoms in the universe. For every atom of silicon there are 40,000 atoms of hydrogen; 3,100 of helium; 21.5 of oxygen; 8.6 of neon; 6.6 of nitrogen; 3.5 of carbon; 0.91 of magnesium; 0.6 of iron; 0.37 of sulfur; 0.15 of argon; 0.09 of aluminum; 0.05 of calcium; 0.04 of sodium; 0.03 of nickel; and 0.01 of phosphorus. Only six of the remaining elements have abundances that lie between a thousandth and a ten-thousandth of the availability of silicon. The others exist only in the minutest traces.

The Earth, when forming, lost most of the volatile gases—unlike, for example, Jupiter. These gases certainly do not exist on Earth in anything like the proportions given in the "cosmic abundance" table. Some of them had, however, combined to form nonvolatile compounds, such as water (hydrogen and oxygen), or the silicates (silicon with oxygen, magnesium, and iron), or the nitrates (compounds of nitrogen and oxygen with metals).

Geologists are able to distinguish between rocks that belong to the surface layer and magmatic rocks extruded from the interior in a molten state. Magma must be similar to the material of the mantle. The oldest extruded rocks are largely composed of olivine. The commonest form of olivine contains about one part of iron to nine parts of magnesium in association with silicon and oxygen. The proportions can vary, however, and it is most likely that the amounts of iron increase farther down in the mantle. Seismic evidence in laboratory tests agrees convincingly with the assumption that the Earth's mantle consists mainly of minerals of the olivine type. Since examination of the meteorites also shows this type of mineral to be dominant, it could give support to the theory that meteorites are fragments of a planet that had consolidated itself by heat and pressure and had acquired an interior similar to that of the Earth.

The Earth's outer core is molten iron. There may be traces of nickel and other "impurities," but nearly pure iron is consistent with physical and chemical data. The explanation of the solid iron inner core is plausibly simple—the heavy iron particles "sedimented" toward the center. That, of course, would also be true of elements heavier than iron, but the proportion of those elements in the Earth is quite inadequate to fill the solid inner core, which seismological studies indicate has a radius of 795 miles (1,280 kilometers).

There have been two estimates of the density at the center of the Earth, one giving it a value of 18 grams per cubic centimeter and the other only 16 grams per cubic centimeter. The oscillations recorded from one earthquake were possible only if the density at the center was 18 grams per cubic centimeter.

The Giant Dynamo

William Gilbert was Queen Elizabeth I's royal physician and as such officiated at her deathbed, but he has achieved immortality for a greater reason: his discovery that the planet Earth is a big magnet. Magnetism had been known long before his time—by the Greeks (the name derives from Magnetes, the inhabitants of Magnesia in Thessaly); by the Chinese; by Muslim traders; and by the Crusaders, navigating the Holy Land. But it was Gilbert who defined the dip and declination of the compass needle and attributed the magnetic forces to the Earth itself, thus explaining the pointing of the needle north and south toward the poles (a term that Petrus de Maricourt, a French Crusader, had applied to the places from which magnetism appeared to originate).

Gilbert's explanation was supported by all observations after his time, including the findings of space research. The question that remained was whether the magnetism originated inside or outside the Earth. A German mathematician and astronomer, Johann Carl Friedrich Gauss, showed that there were two sources, inside and outside.

Subsequent studies have revealed that the internal magnetic field could not originate in either the crust or the solid mantle of the Earth but only in the fluid metallic core. The movement of that fluid core generates electric currents. Eddies in this fluid produce the irregularities that express themselves as observed variations in the Earth's magnetic field at the surface. The fluid motion, however, not only modifies the electric currents but also generates them.

Just as electric current is produced in generating stations by the rotation of metallic wires past each other, the streams of molten metal passing each other in the fluid core "scrape off" each other's electrons and liberate them as flowing currents. The motion of the core can be explained by thermal convection. This is like water in a kettle that is heated from below; the bottom water, as it heats, bubbles to the top. In the case of the Earth this is a globular, rotating kettle.

No one seriously questions the fact that during the last 500 million years (at least) the magnetic field of the Earth has been consistent with its present one. This is borne out by the study of rock magnetism, a fairly recent science that was originated in an attempt to establish a unity between gravitation and magnetism. To perform certain instrumental checks scientists went down into deep mines and there became interested in a magnetic manifestation in rock structures. These take the form of crystals of iron oxide that retain the "remanent magnetization" of the magnetic field to which they were originally exposed. When sedimentary rocks were deposited at various times, iron oxide bore this magnetic imprint. It is as though a compass needle floating in a bowl of water were frozen in the direction of the pole to which it was pointing.

A study of these "pointers" in rocks throughout the world has produced some extraordinary evidence. In successive layers of sedimentary rocks magnetic crystals are differently oriented. There are three possibilities: first, that the axis of the Earth has shifted—even somersaulted—so that the North Pole could have been at the South; second, that the electric currents of the self-excited generator, planet Earth, have repeatedly reversed their direction; and third, that the crust of the Earth has shifted.

By plotting the directions indicated by the remanent magnetization of the iron oxide crystals it is possible to determine where the rocks were in relation to the magnetic poles (which do not coincide with the geographic poles). Judging by those plots, the North Magnetic Pole must have wandered from somewhere in the region where Japan is at present to its current position somewhere in the region of Prince of Wales

Island north of the Canadian mainland. By the same token, from evidence of rocks in the Southern Hemisphere, the South Magnetic Pole must have been a vagrant too. Magnetic rock crystals have been found in the Northern Hemisphere that point to the South Magnetic Pole.

The third possibility—the moving of the Earth's crust—would be quite consistent with this evidence. It would mean that the skin, whether intact or rifted, must have moved so that it was the landmasses containing the rocks that were wandering and not the magnetic poles.

Plate Tectonics

These magnetic crystals, compass needles in the sedimentary rocks, were pointers to the new science of plate tectonics—new in the sense that it brought experts in many branches of science to the rediscovery of the Earth. The idea that continental drift, the movement of the landmasses, would resolve many of the difficulties in the earth sciences belongs to Alfred Wegener, a meteorologist, balloonist, and polar explorer associated with the universities of Marburg, Hamburg, and Graz. Between 1915 and 1929 he published successive editions of his book *The Origin of Continents and Oceans*, refining this theory and rebutting his many critics. Others had noticed before him that the jigsaw of the continents had congruent shapes that could be roughly fitted together into one landmass. He, however, set out to find interlocking evidence about geological formations, fossils, and past climates to demonstrate the breakup of a supercontinent. He envisioned not only fragments drifting apart but also continents colliding to form the upthrust of the Alps and the Himalayas. When he died on the Greenland icecap in 1930, his idea was still derided by many, if not most, of the world's geologists.

All that is now changed, and Wegener has been rehabilitated, with reservations only in terms of information that was not available to him. Crucial to that information was the study of rock magnetism, the operative phase of which began in the 1950s. It originated in the efforts of Patrick Blackett, a British Nobel Prize-winner for his work on cosmic rays and the nature of the atom, to test an abortive theory on the origin of the Earth's magnetism. With extremely sensitive magnetic instruments, recently devised, research teams went into mines throughout the world and studied rock structures. Their work showed that during the past 200 million years of rock formation, Great Britain had rotated clockwise through

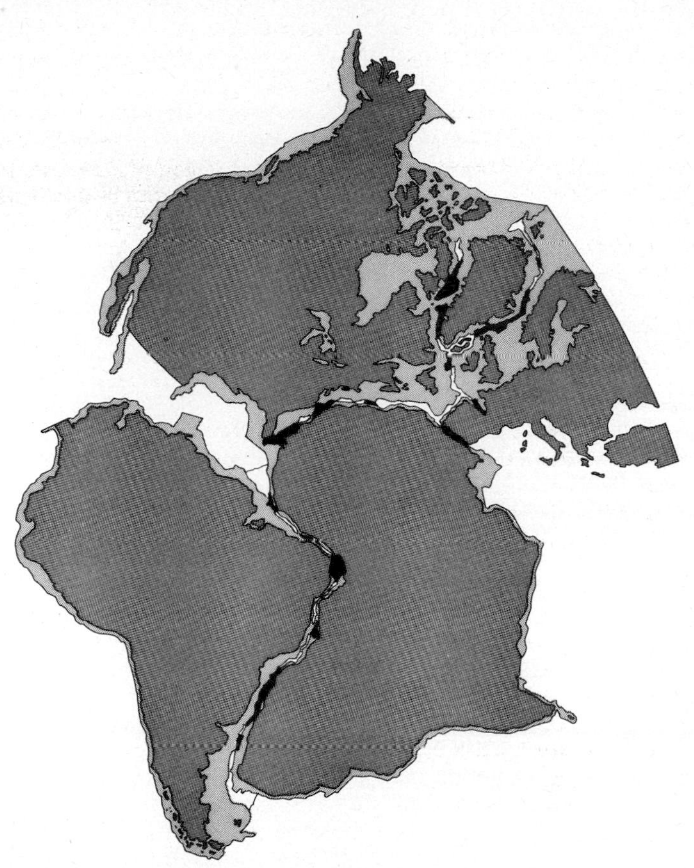

When data concerning the fit of the continents were analyzed, the average mismatch over most of the boundary was no more than a degree. The fit along the continental shelf is shown in light gray; black indicates where land masses, including the shelf, overlap; the white areas are gaps.

about 30 degrees and had also traveled a great distance northward. By 1960 Blackett and his colleagues could maintain, from magnetic data from rocks, that since 500 million years ago the continents have moved north or south at different rates and have also rotated considerably.

In the 1950s oceanographers began discovering more and

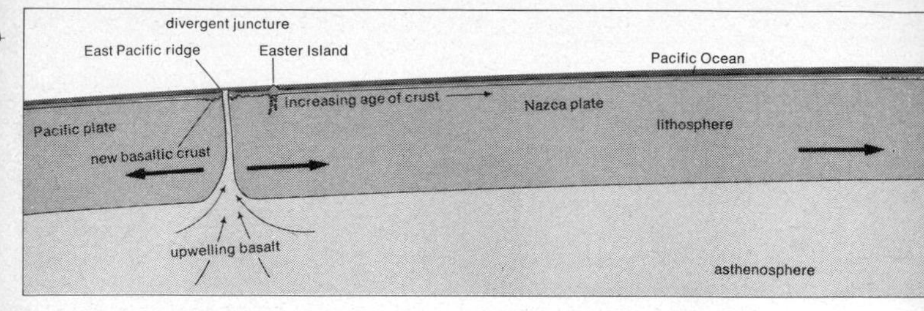

Cross section through the eastern Pacific depicts the birth of new ocean floor along the East Pacific ridge and subduction of the Nazca plate below the South America plate. Diverging Pacific and Nazca plates create a gap that is

more about the seafloor. The mid-Atlantic ridge, a submarine mountain chain, was found to have a deep rift along its summit. It was also recognized as a range that not only stretched the length of the North and South Atlantic but also continued eastward into and across the Pacific. The deep ocean trenches of the Pacific had been plumbed. Theories had been advanced that the ridges were places where hot material was coming up from the interior and that cold material was sinking back into the interior at the trenches. The ocean floor was being renewed and recycled. Again the magnetometers were presenting puzzles. The surveys magnetically scanning the ocean bed were compiling a strange map. The results revealed stripes that were alternately disclosing weaker and stronger magnetism.

In 1963 Fred Vine and Drummond Matthews of Cambridge University maintained that the magnetic stripes could be explained if the rocks of the ocean floor were magnetized in broad bands. This could have happened if the ocean floor at its formation was partly molten. As it cooled, spreading farther away from the originating ridge, the rocks became magnetized by the Earth's prevailing field. Subsequent studies have established that the ocean floor is a huge magnetic tape recorder, telling the story of its own formation and growth. Allan Cox and his colleagues at the U.S. Geological Survey proved from rocks on land, and in confirmation of the oceanic evidence, that there has been a clear-cut pattern of reversals of the Earth's magnetic field over the millions of years.

Since 1969 the *Glomar Challenger*, a U.S. drilling ship, in cruise after cruise has amply confirmed the story told in the ocean's own recording. It has checked the ages of fossils that

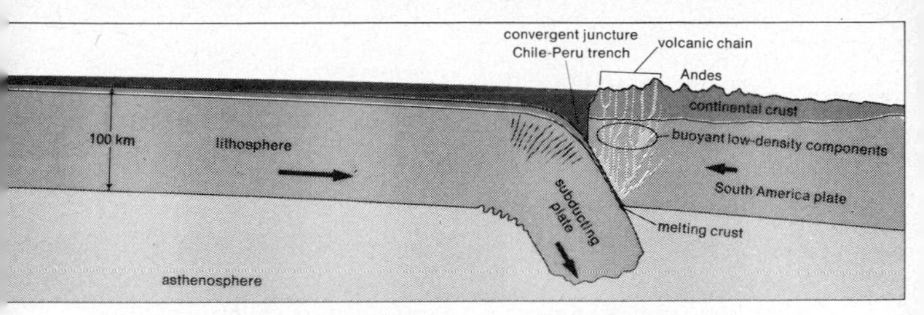

filled by an inflow of liquid basalt from the asthenosphere. As the Nazca plate dives into the asthenosphere and is consumed, certain low-density components rise through the overriding plate to build the Andes mountain chain.

were first accumulated by each portion of the ocean floor, which have shown that the youngest rocks are at the mid-ocean ridge and that the oldest rocks are thousands of miles away. This is evidence that the ocean floor has been spreading and that the continents have been drifting. All this has encouraged the study of, and been explained by, plate tectonics.

Tectonics means construction and has been long used in a geological sense to describe the folding and faulting of the Earth's crust. The rocky surface and the outermost forty miles of the globe are rigid material. If one thinks of the Earth as a chocolate Easter egg that has been cracked, the cracked pieces would be analogous to the "plates." The plates are the rafts of rigid materials that carry the continents. The heat and agitation within the Earth keep the plates shuffling about, jostling each other.

One can get a rough idea of the shape and position of the plates by looking at a map of the earthquake regions. Very little change occurs in the center of the plates. The construction is going on around the edges. No gap can exist between the plates. (There are no forty-mile-deep chasms on the Earth's surface.) One plate can ease itself away from another, but hot rock flows in to caulk the gap. That is what happens at the mid-ocean ridges and in rift valleys. Nor do plates mount each other. If plates are moving toward each other, one of them dips, and the material of that plate passes under the edge of the other to reenter the Earth at quite a steep angle. The "tucking in" of this flap creates an ocean trench. The sinking plate causes deep-seated earthquakes, and the friction it generates causes volcanoes on the far side of the trench.

7.
Fossil Fuels—Their Uses in Engines

The human use of fire; the origin of coal and petroleum and their exploitation; and the use of fossil fuel to power engines—steam, internal combustion, diesel, rotary, jet

Man mastered fire, though it is characteristic of all wild animals (unlike the domesticated cat purring on the hearth rug) that they fear fire. In the primitive world, fire could only be associated with destruction—the lightning flash and the forest fire that it would start; the scorching ravages associated with the white heat of volcanic flow; or the mysterious fires spouting from the ground that we now know as the flames of natural gas or oil seepage. Fire was frightening. Yet somehow, somewhere, early man came to terms with it.

Leaving aside the Greek legend of the sacrilegious Prometheus stealing fire from the gods (a legend common to the mythologies of many people), the rational, but still remarkable, explanation probably lies with lightning striking a tree. Remarkable, because the creature who was to become the fire tender must have had imagination, which is something more than the instincts of animals. Imagination exaggerates the instincts of fear. No form of fire is more frightening than the cosmic force of lightning. Yet someone stayed to observe that lightning started fire and that a spark conveyed that fire. Here was observation and a courage that transcended those of other animals. To take fire and tame it, to confront the lightning and see its fire not as a threat but as an opportunity, was man's first epic victory over nature.

Fire tending must have come first, keeping alive a flame borrowed from nature. Fire making—creating flame at will by flint sparks or wood friction—came later. What fire meant for our primitive ancestors was that they could use it to stave off dangerous animals and to cook food in order to make it more palatable and digestible. It also meant that they could heat their shelters and that, with animal pelts for clothing, they could migrate to otherwise inhospitable climes or survive severe weather. Man's habitat became the globe.

Fuel for the Fire

Fire needed fuel. The obvious fuel was what man could see was combustible, timber and brushwood—and vegetation is still for most of the people of the world their primary source of fire. Some woods were found to be more flammable than others and were used as torches. Then it was discovered that animal fats and vegetable oils were combustible and could be relatively long-lasting as a source of light when surrounding and saturating a fibrous wick. Evidence of candles has been found in the remains of the first Minoan civilization, about 3000 B.C. In Europe during the Middle Ages the luminants of the poor were rushlights, reeds stripped to the pith and dipped in oil. A discovery that led to the whaling industry was that the oil in the head cavity of the sperm whale made a candle of superior illuminating power. Indeed, in 1860 the sperm whale candlelight was established as "candlepower," the measurement of luminance that still persists in the international system of metric units as "candela per square meter."

The Chinese are said to have used coal as long ago as 100 B.C. Marco Polo in his travels there between A.D. 1271 and 1298 recorded the fact that there was a long-established practice of digging out of mountains a black stone that was used for fuel. Theophrastus (371–288 B.C.) referred to a black stone that Greek smiths used instead of charcoal. Excavations in Great Britain have provided evidence that the Romans used coal. Sea coal, coal washed out of seams and collected on the shore or from the sea, was used in Britain and exported to Belgium at least 800 years ago. The first charter to mine coal was granted to Newcastle-upon-Tyne by King Henry III in 1239, and in Scotland the Abbot of Dunfermline was granted a charter in 1291. By 1597 it was "one principal commoditie of this realm of England." By 1661, coal shipped from Newcastle was smothering London in smoke that was "a hellish, dismal cloud which hange perpetually."

Although one thinks of oil as belonging to the twentieth century and the internal combustion engine, its use as fuel dates a long way back. It first revealed itself as seepages and as vents of natural gas, with volatile gases readily flammable. The pillar of cloud by day and the pillar of fire by night described in the Bible could well have been the flare of a vent of natural gas from the oil field of Sinai. The peoples of Persia and Mesopotamia used *naft*, both crude oil and light distillates, for many purposes, including lighting, the dry cleaning of silks, and as detergents for textiles. Alexander the Great used burning petroleum to frighten war elephants. Flammable weapons using pitch and naphtha are of ancient origin and were used as "wildfire" against the Crusaders. "Greek Fire," a secret weapon, was used by the Byzantine Greeks in the siege of Constantinople in A.D. 673. It was catapulted in pots or squirted as a liquid from tubes fixed in the prows of ships. It took fire immediately when wetted and set the sea on fire, enveloping the enemy. It could not be put out with water. Modern research suggests that it was a distilled petroleum fraction with some ingredient, which, like quicklime or phosphide of calcium, gives off heat when mixed with water.

There is evidence that the Chinese were drilling for oil in the third century A.D. They did this in connection with drilling for salt, which they extracted from wells by pumping in water and recovering the salt in solution. The coincidence is geological because salt domes are frequently associated with oil deposits. The domes are upthrusts, attributable to a buoyancy due to the fact that the density of the salt is less than that of the surrounding sedimentary rocks.

Where a salt dome has intruded or uplifted petroleum-bearing sedimentary rocks, the deformation that traps the oil accumulation is commonly over or around the dome. A large fraction of the oil production of the Gulf Coast of Texas and Louisiana is taken from salt dome structures. Oil is also found in salt dome structures in Germany, Romania, the U.S.S.R., West Africa, and various other parts of the world.

There is abundant evidence that early civilizations used petroleum. Bitumen was used as a bonding material in place of mortar. It served as the setting for jewels and mosaics and secured implements and weapons in their handles. It was used to caulk and waterproof boats and to surface roads.

Before coal was mined in Europe, petroleum was being produced on a commercial scale at Baku (now in the Soviet Union). When Marco Polo sojourned in northern Persia in 1271–73, he produced an account of this thriving local industry. Throughout the intervening centuries many travelers brought back reports of this oil field; after at least seven hundred years of production Adolf Hitler in his futile drive for the Caucasus during World War II desperately tried to acquire it to keep his tanks and aircraft going.

The modern history of the petroleum industry dates from the mid-nineteenth century. In Scotland in about 1850 James Young found that oil could be distilled from bituminous coal and shale, and he pioneered various refining processes, but shale oil had to depend on the mining of its raw material and could not compete with the fluid oil that came out of the ground under its own pressure or by pumping. Thus the birth of the oil industry really began with Edwin Drake's successful well at Titusville, Pa., in August of 1859.

Immediately, drilling began throughout the region. About 300 tons of refined oil were marketed the first year. In four years this total had risen to 430,000 tons, and by 1873 it had risen to 1,430,000 tons.

Russia modernized its methods, and in the latter part of the nineteenth century joined the United States as one of the two main sources of the world's oil supplies. At the end of the century the East Indies came into the picture. The Royal Dutch Petroleum Company obtained concessions to develop wells in Sumatra, with the object of selling oil in the Far East. In 1890 the Samuel Brothers, British merchants trading with the Far East, negotiated with the Russians for supplies in bulk to the Orient. The U.S. businessman John D. Rockefeller had already brought 80% of the refining and 90% of the pipelines under the control of the Standard Oil Company. He had set out to control the world market, but after several of his ships carrying oil had met disaster the Suez Canal Company barred oil ships from the Suez Canal. Marcus Samuel, however, had built a special fleet of tankers of a type that reassured the Suez Canal Company, and the ships of the Shell Transport and Trading Company went through the canal with oil from Batum to Singapore and Bangkok. Samuel bought oil properties in Borneo, and eventually the Dutch and British companies amalgamated to become Royal Dutch-Shell. The demand for oil was insatiable because liquid fuel had made possible the internal combustion engine.

Fossil Sunshine Put to Work

It is necessary to remember that coal, oil, and natural gas are the rays of the sun captured by land plants and marine life hundreds of millions of years ago, buried and compressed in the geological pressure cooker, and locked up in the vaults of the lithosphere. Their molecules, however varied the chemical products they produce, are made up of compounds of hydrogen, carbon, and oxygen, with small proportions of nitrogen and sulfur. When the constituent elements of fuel burn, or unite with free oxygen, heat is produced. A fuel is completely burned only when its combustible components are oxidized to the highest degree. In the process a definite quantity of heat is produced, which can be calculated approximately from the chemical composition of a fuel. Thus one pound of carbon in complete combustion to carbon dioxide (CO_2) produces 14,500 British thermal units (BTUs). (A BTU is the amount of heat needed to raise the temperature of one pound of water 1° F at or near 39.2° F.) Only 4,400 BTUs are produced in burning to carbon monoxide (CO), but if this is subsequently burned to CO_2 a balance of 10,000 BTUs is liberated. Hydrogen burns to water vapor, about 62,000 BTUs being produced per pound of hydrogen burned.

The value of a fuel depends primarily upon its potential heat-producing capacity per unit mass (its calorific value), but the calorific intensity, the temperature to which this amount of heat can raise the units of combustion without an excess of air, is also important. Impurities in a fuel affect both its calorific value and its calorific intensity.

The Steam Engine. One way of putting heat to work is to produce steam. The steam engine, a machine for doing mechanical work through the agency of heat, operates by taking in heat at a comparatively high temperature, converting part of the heat into another form of energy, and rejecting the remainder of the heat at a lower temperature. In the case of the steam engine, water is boiled until it becomes vapor and then is superheated into steam.

In a steam engine heat is added to the working substance in a separate vessel, the boiler, where it is vaporized before being admitted to the engine. In the engine the steam expands under pressure, thereby converting part of the heat energy into work. Finally, the remainder of the heat is rejected, either by allowing the steam to escape to the atmosphere, as in locomotives, or by condensing it in a separate apparatus called a condenser, at a comparatively low temperature and pressure. A condenser allows for a greater expansion of the steam, thus increasing the quantity of work obtained.

The fraction of heat added that is converted into work by the engine is called the thermal efficiency of the engine. The addition of a condenser, while increasing the efficiency, also complicates the mechanism. Because it absorbs the heat that the steam rejects during condensation, it requires a supply of cooling water or some equivalent means of keeping the temperature and pressure low. It also requires a pump or other means of removing the condensed substance together with any air that may have leaked into it from the surrounding atmosphere. In the case of large power plants, where fuel economy is important, the advantage of a condenser is so great that virtually all engines employed for this kind of service are of the condensing type.

Given an upper limit of temperature at which heat is taken in, the efficiency attained by the engine is determined by the lowness of the temperature at which heat is rejected. Similarly, when the lower limit of temperature at which heat is rejected is assigned, the engine's efficiency is increased by raising the temperature at which the working substance takes in heat. To secure high efficiency there must be a wide range through which the temperature of the working substance falls, as a consequence of expansion within the engine, from the level of temperature at which heat is received to the level at which heat is rejected. In the steam engine the most efficient performance—the greatest output of work in relation to the heat supplied—is secured by keeping the condenser as cold as the available cooling water will allow and at the same time using a high boiler pressure so that the working substance is very hot while it receives heat in the act of changing from water into steam. For this reason the tendency is always toward higher boiler pressures as the mechanical difficulties of boiler construction and high pressure are overcome.

After conversion into steam the working substance may take in a supplementary supply of heat on its way from the boiler to the engine by passing through a superheater, where its temperature is raised above that of the boiler. A common form of superheater is a group of parallel pipes with their surfaces exposed to the hot gases of the boiler furnace.

Steam engines are classified into two general types according to the manner in which the steam does work during its expansion. In the piston-and-cylinder type, the steam—in a confined space, namely the part of the cylinder behind the piston—enlarges the volume of that space by pushing the piston forward. It does work by exerting a static pressure on the moving piston (the movement of the steam itself is of no significance). In the second, to which belong all kinds of steam turbines, the pressure of the steam is first employed to set the steam itself into motion, forming a jet or group of jets. The momentum of these jets causes work to be done on a moving part of the machine, either by the impulsive action of the jet or jets on revolving vanes or by the reaction on revolving guide blades during the formation of the jets; in some instances a combination of impulse and reaction is employed.

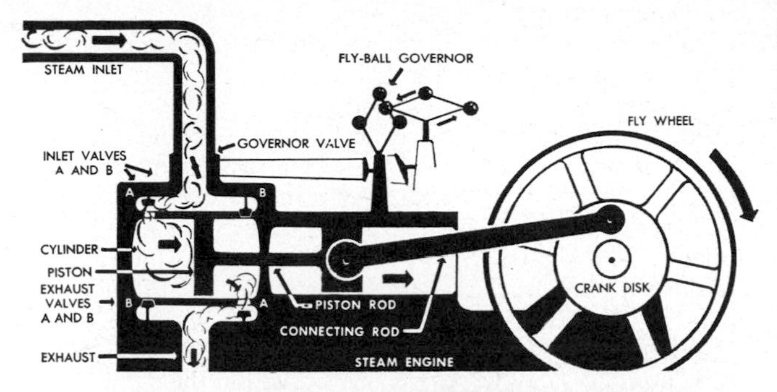

To operate a steam engine, pressurized steam is forced into the cylinder through inlet valve A, driving the piston back and forcing the steam behind the piston out through exhaust valve A. As the A valves close, the B valves open, and the process is repeated to drive the piston forward. This cranks the engine and is repeated again and again to keep the flywheel turning. The flyball governor and valve regulate the supply of steam.

In any turbine the action of the steam is kinetic, in contrast with static action in an engine of the piston type. In both types there is progressive expansion of the steam from the high pressure and relatively small volume at which it is admitted to the low pressure and relatively great volume at which it is discharged. The principle, already stated, that a large range of temperature and pressure between admission and exhaust is essential to efficiency applies equally to both types.

In practice, the turbine has several advantages over the piston-and-cylinder engine. First, it operates much more smoothly because it avoids the reciprocating motion of pistons and the complications associated with them. Second, it is capable of expanding the steam to a much larger volume than could otherwise be achieved, thereby greatly increasing the efficiency of operation. Last, and most important, it can be built to produce tremendous quantities of power with a unit of relatively small size. Because of these advantages the turbine has become the universally accepted method of generating large quantities of power with steam.

Although one tends to think of the steam engine as comparatively recent, it can be traced back 1,800 years. The *Pneumatica* of Hero of Alexandria (about the first century A.D.) described the aeolipile, a primitive steam reaction turbine that consisted of a hollow globe pivoted so that it could turn on a pair of hollow tubes through which steam was supplied from a cauldron. The steam escaped from the globe to the outside air through two bent pipes facing tangentially in opposite directions at the ends of a diameter perpendicular to the axis. The globe revolved by reaction from the escaping steam. Other milestones in the development of the modern engine include the earliest practical model in 1698 by British engineer Thomas Savery, the first successful piston device in 1705 by another Englishman, Thomas Newcomen, and the invention of the separate condenser by the Scottish inventor James Watt in 1765. Sir Charles Parsons, a British engineer, introduced the multistage steam turbine in 1884.

The Internal Combustion Engine. In an internal combustion engine a fuel-air mixture is burned so that hot gaseous products exert force on moving parts of the machine, doing useful work and generating power. The fossil fuel in combustion acts directly on the piston or turbine rotor blades in contrast to the steam engine where it is burned in an external furnace to heat water and produce steam that is then applied

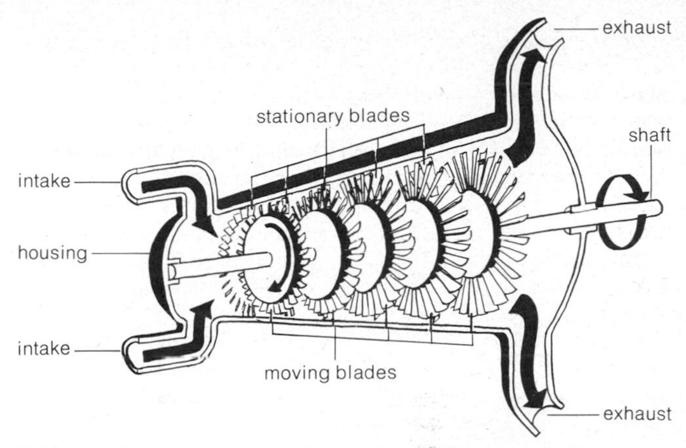

Turbines change the energy of a moving fluid into work. In the steam turbine, the main part, or rotor, is made up of several rows of both stationary and moving blades within a housing. When steam is forced from the intakes to the exhausts the moving blades turn, causing the shaft to rotate.

to the pistons or rotors.

The events leading to the development of the internal combustion engine probably start with the attempt of the French physicist Jean de Hautefeuille in 1677 to pump water by the combustion of gunpowder. The vacuum formed by cooling the hot gaseous products was used to operate the pump. The Dutch physicist Christiaan Huygens in 1680 and the French scientist Denis Papin in 1690 experimented unsuccessfully with similar devices.

The most competent early study of the thermodynamic aspects of the internal combustion engine was that of N. L. Sadi Carnot in France. He published a pamphlet in 1824 in which he outlined not only the fundamental internal combustion engine theory but even that of the diesel engine. Although he did not experiment with actual engines, his *Reflections on the Motive Power of Fire* undoubtedly influenced the thinking of many other experimenters.

In 1860 J. J. Lenoir, a Frenchman, marketed an engine that gave reasonably satisfactory service; it operated on illuminating gas, such as coal gas. It was essentially a converted double-acting steam engine with slide valves to admit gas and

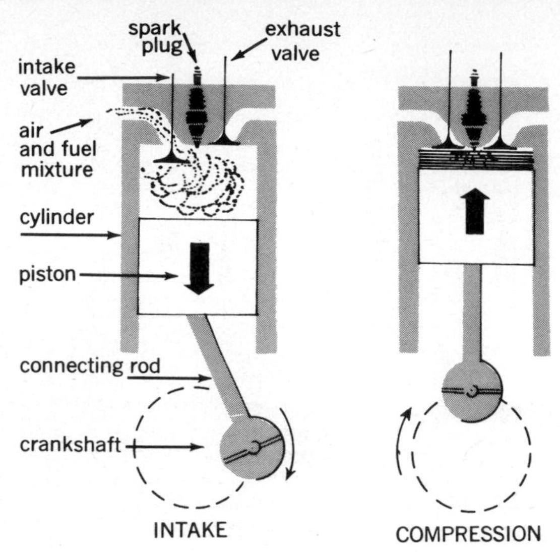

The four-stroke-cycle internal combustion engine makes four strokes in one cycle of operation and repeats them again and again. Intake and exhaust valves in the top of each cylinder open and close automatically to permit the entry of fresh fuel and allow the escape of burned gases. As the piston starts downward on the first or intake stroke, the fuel is drawn

air and to discharge exhaust products. Although it developed little power and utilized only about 4% of the energy in the gas, several hundred were sold over a period of years.

The operating cycle of the modern automobile engine originated with Alphonse Beau de Rochas, who published his theory in Paris in 1862. He stated the conditions necessary for maximum economy as: (1) maximum cylinder volume with minimum cooling surface; (2) maximum rapidity of expansion; (3) maximum ratio of expansion; and (4) maximum pressure of the ignited charge. He described the sequence of operations whereby this economy could be attained as: (1) suction during an entire outstroke of the piston; (2) compression during the following instroke; (3) ignition of the charge at dead center and expansion during the next outstroke (the power stroke); and (4) forcing out of the burned gases during the next instroke.

Although he never built an engine, Beau de Rochas's treatment of factors affecting economy and performance was one

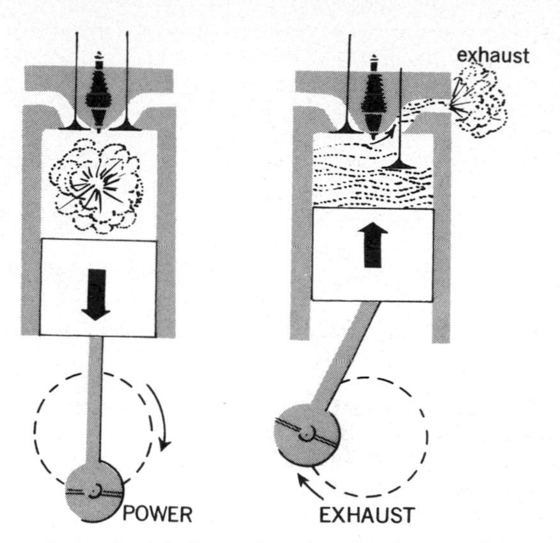

into the cylinder through the intake valve. On the upward or compression stroke the fuel is compressed and an electric spark from the spark plug ignites the fuel, forcing the piston's downward power stroke. The exhaust valve then opens, and burned gases are pushed out of the cylinder by the exhaust stroke.

of the most significant steps in the development of the internal combustion engine. The four-stroke cycle he described is used in modern automobiles. Because his work was entirely theoretical Beau de Rochas has not been given credit for originating the basic principle of the four-stroke-cycle engine. The German technician Nikolaus Otto is most commonly associated with that achievement.

The firm of Otto and Langen started production of a free-piston engine in Deutz, Germany, in 1867. Engines of this type had previously been built, but the Otto and Langen engine was superior in design and workmanship. It employed a vertical cylinder that was open at the top and fitted with a heavy piston, below which was the combustion chamber. The gas and air charge was drawn into the cylinder during the first half of the upstroke. The charge was then ignited electrically and the piston lifted by the explosion. Although the piston tended to continue upward after the pressure on its underside had been reduced to a partial vacuum, atmospheric pressure acting on the top of the piston and the weight of the piston caused it to descend. The motion was transmitted to a shaft and heavy flywheel by a toothed rack extending downward and meshing with a gear on the shaft. This gear turned freely on the shaft during the upward stroke, but a freewheeling clutch engaged on the downward stroke and applied torque to the shaft. The engine was noisy, had little power capacity, and its rack-and-pinion mechanism was unsound mechanically, but it consumed less than half as much fuel per unit of power as did the Lenoir engine and thereby succeeded commercially.

In 1876 the Otto and Langen firm applied the Beau de Rochas principle in the design of the Otto silent engine. This was the first four-stroke-cycle engine employing compression and operating on the principle of the modern automobile engine. In spite of its great weight and low economy, nearly 50,000 engines with a combined capacity of about 200,000 horsepower were reportedly sold in seventeen years; this was followed by the rapid development of a wide variety of engines of the same type. Manufacture of the Otto engine in the United States began in 1878, following the granting of a U.S. patent to Otto in 1877.

Otto's was a horizontal stationary engine working on coal gas. One of his engineers, Gottlieb Daimler, decided that it could be converted into a motor for road vehicles if it could be made small and capable of developing power by virtue of its high speed of rotation. He scrapped the use of illuminating gas and devised a carburetor. This was an ingenious device, consisting of a vessel two-thirds full of gasoline containing an annular float that was attached to a perforated tube. Above this was a small reservoir of mixed air and gasoline from which the engine could draw its explosive mixture. As it did so more air was sucked through the vertical tube and bubbled through the fuel above the float, becoming charged with vapor to replenish the reservoir. Daimler upended the engine from Otto's horizontal position to a vertical one and installed it on a bicycle in 1885.

He had a rival, Karl Benz of Mannheim, Germany. Benz had also started from Otto's engine, but his model was water-cooled. It also included a novel ignition system—a battery, spark coil, and spark plug. The Benz engine was capable of 300 revolutions a minute. Benz also introduced the steering wheel and the driving column. In 1887 his two sons, aged thirteen and fifteen, without parental permission borrowed

Benz's horseless carriage and drove it 84 miles, setting a distance record for the internal combustion engine.

Henry Ford, a machinist in the employ of the Edison Company in Detroit, Mich., saw the possibilities of the gasoline-powered automobile. In manufacturing them he simplified and standardized all parts so that they could be put together on an assembly line and be available as spares all over the country. His Model T, the immortal "Tin Lizzie," was rugged and practically foolproof. A simple kit of tools could keep it going, and a village blacksmith could mend a broken axle.

The Diesel Engine. The diesel engine is an internal combustion engine in which the fuel is sprayed into the cylinders after the air charge has been so highly compressed that it has attained a temperature capable of igniting the fuel. All other internal combustion engines induct and moderately compress an inflammable mixture of air and vaporized fuel. The diesel, or compression-ignition, engine compresses a nonflammable mixture to so small a volume that the injected fuel is fired without a spark.

The engine was developed by Rudolf Diesel, a German engineer born in Paris, France, in 1858. He spent his early years in Paris and later studied engineering at Augsburg and Munich, Germany. Having learned that the thermal efficiency of an internal combustion engine was low, he was inspired to carry out many experiments in order to raise it and thereby reduce the running costs.

He had in mind an improved engine with a cycle of operations approaching that of the ideal cycle described by the French physicist N. L. Sadi Carnot in 1824. He was convinced by results he obtained that he would achieve a higher efficiency if he were to compress the air in the cylinder to a greater extent than was customary at that time. The higher temperature would suffice to ignite the charge introduced into the cylinder, and no fuel ignition equipment would be required. After four years of experimental work he completed in Augsburg his first commercially successful diesel engine (1897). The success of this 25-horsepower, four-stroke, single-cylinder vertical oil engine attracted worldwide attention.

The processes described in Diesel's patent claims made up a cycle differing from the theoretical or ideal cycle followed by existing engines only in the rate of combustion of the fuel-air charge. He proposed to burn the fuel during the first portion of the power stroke of the piston so slowly that no pressure rise occurred. As a means of slowing down combustion to avoid the almost instantaneous explosion of the spark-ignition engine, he proposed to induct air alone into the cylinder, instead of a fuel-air mixture, and compress it so highly that it would attain a temperature sufficient to ignite the fuel that would be gradually sprayed into the combustion chamber during the descent of the piston. This method of igniting the fuel requires that the air be compressed to a pressure of about 500 pounds per square inch, which produces a temperature of about 1,000° F. (537° C).

Diesel presumably thought that any fuel would be suitable for an engine operating in the manner he described. He proposed to build an engine that would burn pulverized coal, the cheapest conceivable fuel, and obtained financial backing for the venture from an organization that had extensive interests in the coal-mining industry. The coal-burning project was soon abandoned, but work continued on an oil-burning engine and, after several failures, the engine ran in 1897.

The fuel economy of Diesel's engine proved to be better than that of any other existing power plant. His engine attracted considerable interest at an industrial exposition in Munich the following year, where Adolphus Busch, a U.S. brewer, saw it. Realizing its possibilities, he purchased a license from Diesel for manufacture and sale in the United States and Canada. A diesel engine built for his company in 1898 was the first to be placed in regular industrial service.

The diesel engine was not rapidly adopted. Until his death in 1913, Diesel insisted that all engines manufactured under his licenses be made to operate with combustion at practically constant pressure, as described in his 1893 patent. This restriction meant that the engines had to run at very low speed. The early engines were so large and heavy in proportion to their power output that they had no application other than as stationary power plants. The original diesel engine weighed 450 pounds per horsepower.

The first marine installation of a diesel engine was made in 1910, very successfully, and the diesel engine became the predominant power plant for submarines during World War I. The first diesel engine that was small enough and light enough for automotive application was built in Germany in 1922 and opened up numerous fields of application that had previously been closed. The high-speed diesel engines do not follow the slow-burning cycle originated by Diesel. Fuel is injected into the cylinder near the end of the compression stroke and burned rapidly, with sharply rising pressure, while

The first diesel engines—large, heavy, and slow—could be used only in ships or stationary power plants. Today's higher speed diesel engines are used to power railroad locomotives, cars and trucks, construction machinery, and snow removal equipment.

the piston is near its dead-center position. Only the compression-ignition and fuel injection of Diesel's original engine are retained in the modern high-speed model.

The diesel engine was the most prevalent power plant for military equipment on the ground and at sea during World War II. After the war the diesel engine became the conventional power plant for all railway purposes in the United States, all heavy construction machinery, a large portion of the trucks and buses operating on highways, and most of the high-powered farm and construction tractors.

The diesel engine is the predominant source of industrial power throughout the world for units up to about 5,000 horsepower, principally because it can burn a low-grade fuel at a lower rate of consumption per horsepower per hour than other internal combustion or steam power plants. It therefore produces cheaper power. A greater fuel saving is effected at partial load than at full load because it is not necessary to throttle the inlet air, as is the case with spark ignition, in order to maintain an inflammable air-fuel mixture. Approximately two-thirds as much fuel is required.

Relatively unrefined fuels can be burned because of the nature of the fuel injection system and the combustion process. This economic advantage has lessened because processes for making gasoline from the heavy constituents of crude oil have increased the value of fuel oils. High-speed diesel fuels now sell for little less than regular gasoline.

Four broad classes of fuels are burned in diesel engines: crude oil, distillates, residuals, and natural or by-product gases. Crude oil can be burned in large slow-speed diesel engines after merely centrifuging it to remove sand and water. The distillates, called gas oils by the petroleum industry, are taken from crude oil after the volatile fractions used in gasoline are removed. They range from light fuel oils, differing little from kerosene, to the heavier distillates that have boiling points within the approximate range of 350° to 650° F (175° to 345° C). Residuals include all of the ingredients of crude oil remaining after the gasoline and distillate fuels are removed. They are cheaper than the distillates but cannot be burned in high-speed diesels and may cause operating difficulties in larger engines. Natural gas is burned in practically all large, stationary, power plant diesel engines. The resistance of gases to self-ignition necessitates pilot injection of fuel oil, but cheaper power is produced than with any other fuel.

The Rotary Engine. Continuous rotation is superior to

reciprocating motion. The fundamental principle of the gas turbine was known before that of any other type of heat engine. Hero of Alexandria had a perfectly plausible turbine nearly two thousand years ago. Nevertheless, practical development of the gas turbine did not take place until the twentieth century.

The gas turbine is a simple power plant. A compressor supplies air at about three to six times atmospheric pressure to a combustion chamber into which fuel is sprayed, maintaining continuous combustion. The resulting volume of high-temperature gaseous products then expands through the turbine unit to atmospheric pressure. The compressor and turbine rotors may be on the same shaft. The excess power developed in the turbine over that required to turn the compressor rotor is delivered to a generator, propellor, or other load. In the free-piston engine the pistons are not connected to a crankshaft, as in a conventional engine, but can transmit their power to a turbine instead. Originally built as an air compressor, the engine was first used extensively by the Germans in World War II for launching torpedoes.

In spite of the theoretical advantages of rotary drive, the automobile industry was reluctant to develop it. In the early 1950s a rotary engine for motor cars was invented by Felix Wankel in West Germany. The first attempts at commercial versions were unsatisfactory, but the Japanese saw possibilities, improved the design, and manufactured such cars in volume. In 1970 General Motors agreed to pay $50 million for the U.S. patent rights. The Wankel is a simple, small, lightweight engine. Engineers expected it to be cheaper to build than the motor in today's cars—easier to service, quieter, and with less vibration. But the feature attracting the most attention was its potential in the fight for cleaner air. The present internal combustion engine emits carbon monoxide, unburned hydrocarbons, and oxides of nitrogen. The Wankel engine does too, but because it is smaller there is more room under the hood for antiemission devices, and its venting system makes emissions easier to handle.

In the piston-driven car gasoline and air are mixed in a carburetor. The resulting fumes are then pumped into a combustion chamber where they are compressed by pistons and ignited by an electric spark. The gaseous mixture then explodes, forcing the four or six or eight pistons to keep pumping up and down, thereby supplying the power that eventually turns the wheels.

In the Wankel the air-fuel mixture is compressed and ignited as in the piston engine, and expansion from the burning fuel provides the force that eventually turns the wheels. There are no pistons moving up and down. Instead, a triangular-shaped rotor revolves around a shaft in an eccentric path within the working chamber, and the spinning shaft sends power to the wheels. The tips of the rotor continually touch the chamber surface as it spins, forming continually changing pockets of fuel. Fuel enters an intake port and is compressed as the rotor swings around, squeezed in the narrowing space between the chamber wall and the rotor face. A spark plug ignites the fuel, and the gases expanding against the rotor face drive it around and thus turn the shaft. As the cycle continues, the rotor swings past an open exhaust port and the spent gases leave the chamber. A Wankel usually has fewer than half the moving parts of a conventional engine and takes up less than one-fourth the space.

A drawback of the rotary engine in an era of rising prices for fuel is that it uses more gasoline per mile than a piston model. This is being improved by new designs, however.

The Jet Engine. Jet propulsion—whether the term is applied to a stick rocket at a carnival, a space vehicle on the way to the moon, or a jumbo airliner—is based on the reaction of a body to the rearward thrust of a jet of gas. The physical principle was laid down by Isaac Newton in his *Philosophiae Naturalis Principia Mathematica*, in which his Third Law of Motion requires that for every action there is an equal and opposite reaction.

A simple example of reaction propulsion is the brief flight of a toy balloon after it has been filled with air and released. The forward motion of the balloon results from the rearward expulsion of air from the balloon. The thrust does not result, as is sometimes erroneously presumed, from the jet pushing against the surrounding air. The force imparted by the jet is equivalent to the product of the mass flow of the air and the velocity of the jet. In the case of a chemical-propellant rocket the jet is composed of the gaseous combustion products of the propellant mixture, which is burned inside a thrust chamber and ejected at supersonic velocity through a nozzle. The gas velocity in a properly designed converging-diverging nozzle is sonic at the converging throat (most narrow) portion of the nozzle and becomes supersonic as it travels through the diverging (exhaust) end of the nozzle.

A jet engine is distinguished from a rocket by the fact that

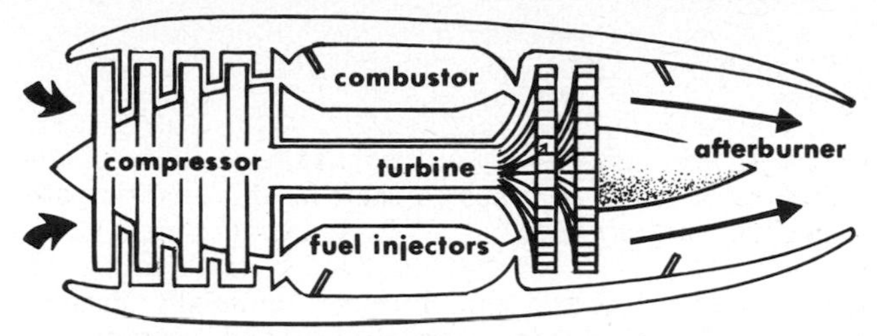

A turbojet engine develops power by drawing in and compressing a large quantity of air, which is then mixed with fuel for combustion. The resulting gases expand and produce thrust. In some turbojets, additional power is developed by introducing fuel to the escaping excess air in an afterburner.

it is an air-breathing device, whereas rocket fuel is self-sufficient and self-contained. The typical jet engine is shaped like a cigar. Air is drawn in, compressed, heated by combustion with a fuel (kerosene), and expelled with the resultant gases with sufficient force so that there is a substantial thrust in the opposite direction.

The idea of rocket propulsion had been around a long time before Flight Lieut. Frank Whittle of Great Britain registered a patent for a jet engine in 1930. At the same time, completely independently, Hans von Ohain in Germany had been proceeding on much the same lines. His version was airborne on Aug. 27, 1939, as the Heinkel He 178. Whittle's engine made its first flight on May 15, 1941.

Whittle's and von Ohain's devices were both what have come to be called turbojet engines. The basic idea of this engine is simple. Air taken in from the atmosphere is compressed to 3 to 12 times its original pressure in a centrifugal or axial compressor. Sufficient fuel is added to the air and burned to raise the temperature of the fluid mixture to between 1,200° F and 2,000° F (655° C and 1090° C). The resulting hot air is passed through a turbine, the sole function of which is to drive the compressor. If the turbine and compressor are highly efficient, the pressure at the turbine discharge will be nearly twice atmospheric pressure. This excess pressure is then used in a propelling nozzle to produce a high-velocity stream of gas and, therefore, a thrust.

The most important advantage of the jet engine is that it is extraordinarily light as compared to an engine-propeller combination having a similar thrust at cruising speed. This advantage makes it possible to use much more powerful engines in jet planes. In addition, the elimination of the propeller makes possible a simpler aerodynamically cleaner (more streamlined) air frame. Furthermore, the propulsive efficiency of a propeller is low at high forward speeds (above 500 miles per hour), while the propulsive efficiency of a jet is particularly good at very high speed. All these advantages taken together made it possible to build turbojet-powered aircraft having a speed advantage over their engine-propeller counterparts of more than five to one.

The German V-1 bombing weapon used in World War II was a pilotless airplane provided with an intermittent jet-propulsion device called a pulse jet. In this device air entered at the front through a nonreturn valve and passed into a combustion chamber, where it was mixed with gasoline and ignited. Because return of the air was prevented by the valve, the resulting combustion raised the pressure in the chamber and forced the products of combustion to flow out of the nozzle at the rear. Inertia of the flowing gas in the nozzle caused the gas to continue to flow until the pressure in the combustion chamber was reduced below atmospheric pressure, at which time the nonreturn valve opened and allowed a new charge to enter. Because it was found that residual fire from one explosion would ignite the next, an ignition device was necessary only for the first explosion.

The pulse jet was simple, cheap, and effective for its intended purpose. It has never become a propulsive device for manned aircraft because of its inherently high fuel consumption and because it produces an almost intolerable noise.

The simplest of all aircraft propulsive devices is perhaps the ramjet. It is essentially a turbojet in which the rotating machinery has been omitted. It functions only at high airplane speed. Air entering the front of the device at flight speed is slowed to a low velocity (relative to the airplane velocity) by providing a properly shaped inlet passage. If this process is done efficiently, the pressure will be raised. Fuel is added and burning takes place at the higher pressure. The hot products of combustion are reexpanded to atmospheric pressure and flow out at a high velocity through a rearward-facing nozzle. Because of the higher temperature of the gas after combustion the velocity at the nozzle exit is higher than the flight speed; therefore, a thrust is produced. The ramjet can be an efficient device only at extremely high speed. It can produce no thrust at standstill and little at a low forward speed.

The turbofan jet is much like the turbojet except that a fan is mounted on the turbine shaft just inside the front of the engine housing. The fan helps draw in additional air, increasing both the power and the thrust of the engine.

Early turbojet-powered transports had marginal takeoff performance, necessitating a considerable reduction of payload on long flights. This stimulated the development of the turbofan engine, an effective means of improving takeoff thrust without large increases in engine weight. By adding to a turbojet a fan driven by one or more turbine stages the thrust of a given engine can be increased. The increase is particularly large at the low speeds corresponding to takeoff and climb conditions, but in addition the increase extends into the range of cruising speeds without an increase in fuel consumption. The turbofan, therefore, provides just what a conventional transport airplane needs—a substantial excess of thrust at takeoff and low fuel consumption at cruising speed.

8.
Electricity— Its Behavior as Power

The behavior of electricity in various types of matter; its use as power; and its applications in transistors, sensors, lasers, luminescence, xerography, and fuel cells

When British physicist Sir J. J. Thomson established the existence of the electron in 1897, an innocent asked him, "What does it look like?" He replied, "It is a red-nosed pixy chasing other pixies around."

We still have not seen the pixy, but we have seen its tracks and we know a great deal about its nature and behavior. It is a particle with a negative charge. It weighs only 1/1,837 as much as an atom of the lightest element, hydrogen. We know accurately the number of electrons in a wire. If the wire is made of copper, each copper atom will have 29 electrons circling round the nucleus and there are 1.385×10^{24} (10,000,000,000,000,000,000,000,000) atoms per cubic inch. Most of those electrons take no part in carrying current through the copper wire but stay with the atoms where they are, while one or two electrons from each atom move freely through the metal as carriers, or "free electrons." Even so there are so many carriers that they do not have to flow very fast to carry an ordinary current. The individual electrons are in a great tizzy, moving fast but at random, and the systematic drift that constitutes the current is traveling at much less than an inch per second. The waves in the flow of electrons, however, make telegraphic speeds.

In the atom electrons orbit around the nucleus. In the simplest case, hydrogen, one electron circles around one proton, which is positively charged. A copper atom contains 29 positively charged protons and 29 negatively charged electrons. Uranium has 298 of each so that the orbits of the electrons are like a ball of knitting wool.

Electron Behavior in Matter.

Protons repel protons, and electrons repel electrons. Electrostatically, a proton exactly balances an electron in the sense that a proton attracts an electron exactly as strongly as another electron at the same distance repels it. To recognize electrons as electricity, one must consider their behavior in various states of matter.

Currents in Molten Salts and Water Solutions. When a salt is heated to the melting point, the thermal agitation is increased to such an extent that the electrostatic forces are unable to keep atoms from exchanging places and wandering around. Charged atoms are then called ions (Greek for "wanderers"). Their wanderings can be given preferred directions (positive or negative) by the use of electrically charged metal plates called electrodes.

A flow of ions of either kind or of both in opposite directions is equivalent to a flow of electrons, but it introduces complications not found in wires. If an end of a copper wire is connected to an iron wire, electrons flowing through the copper can flow right on through the iron because they are just as much at home in one as in the other. But when chlorine ions flowing through molten sodium chloride come to electrode P, the positive of the two electrodes, they cannot flow on through it. There, the current can be carried only by its free electrons. If electrons are to flow steadily through the surface of electrode P, therefore, they must be carried to this surface as the excess electrons in chlorine ions and then away from it as free electrons within the metal. That is, at the electrode surface the electrons must take leave of the chlorine ions. Then these ions, losing their charges, become electrically neutral atoms and join themselves into pairs as molecules of chlorine gas, which accumulate on the surface of electrode P.

In this reaction some of the electrons may not flow through the electrode surface but only from it, starting from surface atoms of the metal. Such atoms, left deficient in electrons, become ions and dissolve into the molten salt. The more reactive the metal, the more the surface reaction takes this form. With any metal, however, there is at least one reaction at this surface, whereas there is none at a junction between two wires.

At the surface of the negative electrode E there is another reaction. Within electrode E its free electrons are flowing toward the surface to meet sodium ions attracted toward it. These ions, being deficient in electrons, capture them at the surface and accumulate as neutral metallic sodium. This is, in

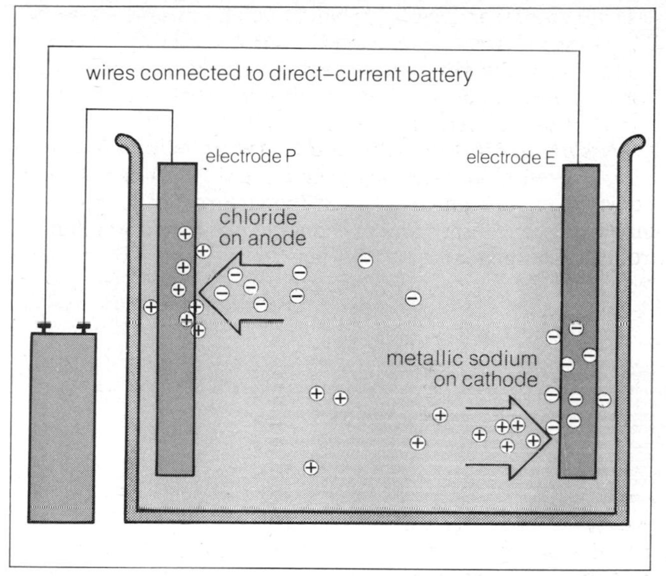

Electrolysis takes place when a pair of electrodes (electrode P and electrode E) are immersed in molten salt and are then connected with a source of direct electric current. The positively charged sodium ions are attracted to the negatively charged electrode E, where each acquires an electron, loses the positive charge, and becomes an atom of metallic sodium. Similarly, the negatively charged chlorine ions collect at the positively charged electrode P, release their extra electrons, and become chlorine atoms. This transfer of ions permits the electric current to flow.

fact, one of the easiest ways to reduce this highly reactive element to metallic form.

The name given to such conduction, as by this molten salt, is "electrolysis." The molten salt is an electrolyte, and the current electrolyzes it. Water solutions of such salts also are electrolytes, and so are solutions in liquid ammonia and some other solvents. In them, the reactions are often essentially like those in molten salts, but with further complications. For example, with sodium chloride in water and with iron electrodes, the reaction products at electrode P include chlorine gas and ferrous chloride, but at electrode E, because metallic sodium would react instantly with water, the products are hydrogen gas and sodium hydroxide.

Currents in Gases. Another sort of conduction involving ions occurs in gases; for example, through the air in an electric spark or through neon in a neon sign. These currents obey laws quite different from those of electrolysis.

Gases are not habitually ionized like electrolytes. Cool air is an excellent insulator. In an electric spark, however, the air is a fairly good conductor because the high temperature causes violent impacts between molecules. These impacts drive electrons out of some of the molecules, making them into ions. Cool neon, likewise, is a good insulator until violent impacts ionize it.

In both cases a few ions are present before the current starts. Even these few are the result of something locally violent, such as a cosmic ray or a stray atom of some radioactive element. Ordinarily there are too few of them to conduct more than a bare trace of current. To start a spark, they must be made to increase their number greatly. Each of these few ions must be driven to exceptionally violent impacts. The spark starts, therefore, only when the electrostatic field is much stronger than it will need to be later when the air has become hot. In the neon sign there is less difference of this sort.

A second important characteristic of a gas ion is that, whether it is a molecule deprived of an electron or a molecule to which a free electron has attached itself, when it reaches an electrode and is neutralized it does not stay there. In this way gas ions are quite different from ions in electrolytes.

Historically, the study of currents in gases has been important; it has led, for example, to the discovery of X rays and of the electron. Practically, also, it is important to certain persons, such as designers of neon tubes and fluorescent lamps and power-line engineers, who have to design against sparking and even against quieter losses of power.

Electrons in Wires. Just as ions are driven through a molten salt by electrostatic repulsions and attractions, so are electrons in wires. Sometimes, to be sure, these electrons are subject also to other forces, such as electromagnetic forces acting in the moving wires of a motor or a generator; but with or without such forces, electrostatic forces always are present and are fundamentally important.

Their importance arises from the fact that it is through these forces that power is transmitted. When a lamp filament glows, the power to keep it hot is delivered to it by the electrostatic forces on its electrons, that is, by repulsion on them by one end of the filament, which has an excess of electrons, or attraction by the other end, which has a deficiency of them, or both. If the charges on the ends of the filament were not maintained by the continual pumping of electrons against electrostatic forces at the source of the power (generator, battery, or the like), the current through the lamp would draw off the excess electrons and fill up the deficiency in less than a millionth of a second. These charges do not contain any great reserve of energy, but they are of basic importance in understanding how the lamp, or any other electrical appliance, really works.

Force and Direction. If the electrons in two parallel wires flow oppositely, there is repulsion. This relation of force to direction made it necessary to have some way to assign a conventional direction to any current, long before anyone knew in which direction anything really flowed. Even before the French physicist André Ampère discovered these forces, Benjamin Franklin and others had found a need for algebraic signs for charges, to use in equations about their forces.

For a long time it had been assumed that charges of either kind could flow in metals, but Franklin pointed out that this assumption was unnecessary. His one-fluid theory was essentially that electrons flowed in metals while protons had fixed positions. Unfortunately he was wrong in his description of electric current. When, according to Franklin's view, a wire is said to be carrying a current from left to right, the electrons in it are really flowing from right to left.

Strangely enough, this false convention caused no serious inconvenience for many decades, and it still causes none in most of the branches of electrical engineering dealing with motors, generators, power transmission systems, and electric lighting. In such devices electrons simply circulate around closed metallic circuits, remaining within the metal; in these circumstances, if all currents are reversed, all outward manifestations of force, power, heat, and light are unchanged. With vacuum tubes, on the other hand, the situation is quite different. In a tube of almost any type used in radio and other communications systems, there is at least one electrode that can be heated to enable electrons to escape from it into the vacuum, while other electrodes, kept relatively cool, can only receive electrons from the vacuum and cannot emit them. Through such a tube electrons can flow from a hot electrode to a cold one, but not the other way. In the wires connecting the tube to other things, therefore, the real direction of a flow must be recognized.

In the wire connected to the hot electrode, electrons can flow only toward the tube, whereas in the other wire they can flow only away. Consequently, when diagrams based on the false convention show "currents" opposite to these flows, the convention is an endless source of confusion to students. Even to an expert, the need for thinking always the opposite of what is said, and guarding against wrong algebraic signs, is a perennial waste of mental effort.

Much of this waste, though admittedly not all, can be prevented by dropping the term "current" and speaking of "electron flow" whenever there is any consideration of its direction. This procedure indeed comes near to eliminating the trouble entirely. What remains of it is a set of + and − signs, some of them on electrical measuring instruments and the rest in equations referring to electrostatic charges.

These signs are present because of the natural guess that, if a "current" is running "into" a body, this body is receiving something, just as a basin receives water when a stream runs into it. So a body is said to be gaining a positive electrostatic charge, when what it is really doing is losing electrons. The conventional + signs, therefore, go with deficiences of electrons, while the − signs go with excesses.

Electric Power

Almost all uses of electricity are also uses of power. This power may be a thousand kilowatts for an electric locomotive; it may be a hundred watts for a lamp; or it may be only the minute fraction of one watt that a radio receiver picks out of "the air." Whether it is measured in kilowatts or in megawatts, or is simply read in watts from the trademark on a lamp, the power and its basic unit, the watt, are central features in almost any question about the use of electricity.

Nevertheless, the watt is not really an electrical unit. Power was used in mechanical engineering long before electricity seemed to have any possibility of use. Any unit of power, therefore, belongs to the science of mechanics; and so the watt is appropriately named for James Watt (1736–1819), the inventor of the first good steam engine. The definition of the watt naturally is based on the technical definition of power. This, in turn, is based on work. The words *power* and *work* obviously were suggested by their nontechnical meanings,

All bodies have a weight or downward force of gravity that is proportional to their mass. The amount of work done by gravity is figured by multiplying the weight of the body by the height of descent.

related to muscular effort, but they have been given technical restrictions that make them more definite in meaning, as well as more useful.

Gravity does work on any body that moves downward. The amount of work is found by multiplying the force of gravity (the weight of the body) by the height through which it descends. If the descent is oblique, only its vertical component is used in computing the work. In general, as in this case, the work done by any force, acting on any moving body, is a product of force and distance. If a 200-pound man, for example, goes from one floor to another ten feet lower, the work done by gravity is called 2,000 foot-pounds. This compound unit, the foot-pound, is a prototype for many units formed and defined mathematically.

Power is defined as the rate at which work is being done. The power exerted by gravity on the 200-pound man, for example, depends on his rate of descent. If his ten-foot descent is accomplished in five seconds, the power is 2,000 foot-pounds divided by five seconds, or 400 foot-pounds per second. Any unit of power like the foot-pound per second is a compound unit, whose definition includes division as well as multiplication.

The watt is the unit of this type most suitable for use in relation to electricity. Since the ampere (the unit for measuring current) is defined in terms of the newton and the meter, these are the best units of force and distance for use in the watt. (The newton is the unit of force that produces an acceleration of one meter per second per second in a mass of one kilogram.) The time unit to use with them is suggested by the definition of the newton, which involves the second. Therefore, the watt is defined as one newton-meter per second.

To emphasize the strictly mechanical character of the watt, we may express that 200-pound man's weight approximately as 900 newtons, and the ten feet as three meters. The power released by gravity is therefore:

$$P = \frac{900 \text{ newtons} \times 3\text{m}}{5 \text{ sec}} = \frac{540 \text{ newton-meters.}}{\text{second}}$$

The word *watt* is simply an abbreviation for this last fraction. That is $1 \text{ w} = \frac{1 \text{ newton-meter,}}{\text{second}}$

so this power is 540 watts.

This algebraic form of the definition of the watt has the advantage of showing at a glance how the power depends on force, distance, and time. The positions of the newton and meter above the fraction line reveal that if either the force or the distance is increased, other things remaining the same, the power is increased. The position of the second below the line indicates that increasing the time, for the same force and distance, reduces the power. Most electrical units, like this purely mechanical one, are fractions formed out of previously defined units.

Kilowatts and Kilowatt-hours. When we pay an electric light bill we do not pay for a number of kilowatts, even though we may carelessly say we do. This is similar to a sign on the highway saying "Speed 25 miles"; 25 miles is not a speed. The common misuse of "kilowatt" is like this, but in reverse. The kilowatt, being 1,000 newton-meters per second, is not like the mile but like the mile per hour; it is a unit of the rate of doing work rather than of the total work done. What we pay for, each month, is the total work done. If four 100-watt lights are kept turned on for 200 hours altogether, the work is

$$4 \times 100 \text{ w} \times 200 \text{ hr} = 80{,}000 \text{ w-hr} = 80 \text{ kw-hr}.$$

It is kilowatt-hours, therefore, rather than kilowatts for which we pay. One kilowatt-hour is

$$1{,}000 \text{ w} \times 1 \text{ hr} = \frac{1{,}000 \text{ newton-m}}{\text{second}} \times 3{,}600 \text{ sec} = 3{,}600{,}000$$

newton-m.

By human standards this is a prodigious amount of work to do for a few cents. One newton is about 0.22 pounds, and one meter is about 3.3 feet, and so the kilowatt-hour is easily found to be about 2,600,000 foot-pounds. This is roughly the work done by thirty men climbing the Washington Monument via the staircase, regardless of whether they do it in an hour or in ten minutes.

With the watt, the kilowatt, and the kilowatt-hour defined in purely mechanical terms, they may be applied to questions about electricity. The watt, as already noted, is mentioned in the trademark on an incandescent lamp or light bulb; and along with it is another unit, the volt. If the trademark states "100 w 120 v," this means that the light will take 100 watts when connected to a pair of wires for which one may say "120 volts." The "100 w" refers to power exerted on the electrons flowing through the filament of the bulb. This power is exerted by the electrostatic forces described above. These forces are what must comply with the specification "120 v."

One can move a light bulb from one fixture to another in the same house or even into another house, usually without raising the question of whether the electrostatic forces in the new location will comply with this specification. Usually, indeed almost always, they do. But if the question is raised, this compliance becomes evidence that there must be a general law compelling the charges on the wires to distribute themselves so as to produce the specified forces.

This law, actually, is the basic law of repulsions and attractions between electrons and protons. These forces are much like forces of gravity. Between any two small particles a force of either type acts directly along the line from one particle to the other. Its strength depends only on the distance between them, and it is not affected by the presence or absence of any third particle. These similarities to gravity enable us to carry many laws of gravity over into electrostatics. Among these are laws about work and power.

If water flows from a millpond to a pool below the dam, the work done on it by gravity is the same, regardless of whether it goes through the mill and turns most of this power to good account or merely becomes the proverbially useless, though slightly heated, "water over the dam." Relating this law to electrostatics, we may apply it to electrons flowing through different lights or motors. Starting on one terminal of the big double-pole switch in the electric feeder lines of a house (usually in a box on the wall), one electron may find its way through a light to the other switch terminal, and another electron may find its way through a motor. These ways may look different, but the work done by electrostatic forces on one of these electrons is exactly equal to the work done on the other. This law is especially important in its application to the operation of household electrical appliances.

If the number of electrons passing through the light and the motor in a second are equal, the rates of doing work—that is, the amounts of power—are also equal. If (as is more likely) five times as many electrons go through the motor as through the light, they receive five times as much power from the electrostatic forces.

In short, when electrons flow from any one point (such as one terminal of that big switch) to another point (such as its other terminal), the power delivered to them by electrostatic forces is proportional to the number of the electrons and does not depend on the paths of the electrons or on the type of conductor through which they flow.

Potential Difference and the Volt. Any law of proportionality, such as that just described, can be expressed in a statement that the value of some specially defined fraction stays constant. In this case, the fraction is power divided by current. The name of this fraction is "potential difference."

Consider a light bulb marked "100 w 120 v" and compare it with a slightly larger one marked "150 w 120 v" and a still larger one marked "200 w 120 v." Suppose each of these bulbs, alone, is connected in series with an ammeter, to measure its current, and that the ends of the series are connected to a pair of wires conforming to the specification "120 v." Then the ammeters will read about 0.83 amp, 1.25 amp, and 1.67 amp, respectively. So the fraction, which stays constant, may be calculated as

$$\frac{100\text{ w}}{0.83\text{ amp}},\ \frac{150\text{ w}}{1.25\text{ amp}},\ \text{or}\ \frac{200\text{ w}}{1.67\text{ amp}}.$$

With reasonable allowance for instrumental errors these values for the fraction are equal. Each is practically $\frac{120\text{ w}}{1\text{ amp}}$, and this is what is meant by "120 v."

The word *volt*, therefore, like the word *watt*, is simply a short name for a compound unit, defined as a fraction; that is, by definition,

$$1\text{ volt} = \frac{1\text{ watt}}{\text{ampere}}.$$

Its derivation is from the name of Alessandro Volta, the Italian physicist who invented the battery (described by him in 1800), because this was the first device for maintaining steady potential differences between terminals of conductors.

Transmission of Power. Electrons are subject to forces of several common types other than electrostatic. Among these are chemical forces in batteries and electromotive forces in generators, transformers, and even motors. Indeed, a battery, a generator, or a transformer is built and used primarily as a sort of electrical pump to drive electrons against electrostatic forces, somewhat as a water pump drives water against gravity.

Gravity makes such use of water possible because it repays, for example, in the mill, the work done against it in the pump. Gravity can accept and repay work in this way because it is present in both the pump and the mill. In exactly the same way the electrostatic field—the region where electrostatic forces can be found—can accept and repay work done on electrons because it covers the whole apparatus: generator, transmission line and motor, and light or other receiver of electric power. In short, the transmission of power by electricity is accomplished by driving electrons against electrostatic forces in one place and letting these forces drive electrons in another place.

Electromotive forces cannot be measured in force units, such as pounds or newtons, but like potential differences they can be measured in volts. Because of this the colloquialism "voltage" is often used to cover both potential differences and electromotive forces. In using it, however, care must be taken not to confuse these quantities.

An automobile battery, for instance, usually has an electromotive force of about 12 volts. That is, when current flows through it in obedience to its chemical forces, power is exerted at the rate of 12 watts per ampere; as an example, 60 watts for headlights taking 5 amperes. A little of this power is lost in driving ions through the liquid of the battery. If this loss is 2 watts, only 58 of the 60 watts are available for work against electrostatic forces, and so the charges adjust themselves accordingly. That is, while the electromotive force within the battery is 12 volts, the potential difference between its terminals is only 11.6 volts; thus, these "voltages" differ in strength as well as in nature.

The whole duty of a generator or a battery is to drive electrons against electrostatic forces so that elsewhere in the same field the electrostatic forces can do the driving and deliver the power. Water could be used like electrons, but there is one important difference, although it is more a difference in degree than in kind. If water runs from a pump to a mill through a canal and the pump stops, the mill need not stop until it has drained the canal. Likewise, when a switch is turned off, electric lights need not go out until they have drained the wires that have excess electrons; however, they do go out very quickly. Wires, evidently, differ from canals in being drained rapidly by the appliances they feed. Even larger electrical systems have this characteristic; a lightning flash goes out as "quick as lightning."

To maintain a steady current in a light bulb, therefore, the wires must be treated almost as if they had no capacity as reservoirs for electrons at all. The wire that feeds electrons to the bulb must have its supply of electrons replenished continuously by the source of power, and the wire that receives electrons from the bulb must likewise have them taken away.

Circuits are of many types, some amazingly complicated. In many, notably those of radio systems where charges do change considerably in fractions of a millionth of a second, the currents involved in draining and rebuilding the charges are important. Most household and industrial circuits contain not batteries but transformers. One essential difference, however, should be noted. A battery pushes its electrons in the same direction all the time, but a transformer cannot do this for more than a small fraction of a second. Consequently, transformer circuits must carry alternating currents; that is, currents whose directions alternate many times every second.

Wireless

Electricity is not confined to wires. As we have seen, it travels through space, and by means of radio astronomy we can pick up signals from the immensity of the expanding universe, recording the birth cries and death gasps of stars. Or we can pick up the signals pulsed by the human brain. Whether the signals are naturally propagated by the stars or by the brain cells or are made by radio transmitters, they have to be detected and reassembled as telegraphic messages, voicecasts, or television images. They may have traveled for millions of light-years or for milli-microseconds as in the computer, but their energy has to be converted into audible or visible versions adapted to the human ear or eye.

Among the pioneers in this field was Joseph Henry, who in the mid-nineteenth century stretched parallel wires across the campus of Princeton University in front of Nassau Hall and detected signals that passed between them. In the late 1880s the German physicist Heinrich Hertz proved that radio waves and light waves differed in wavelength only. Hertz not only proved the existence of radio waves but showed that they could be reflected from solid objects as in radar. In 1896 both Guglielmo Marconi, an Italian physicist, and Aleksandr Popov, a Russian, showed how radio signals could be sent and received.

In his work on the electric light, Thomas Edison had produced a filament by inserting a loop of carbonized bamboo into a glass bulb evacuated of air (long since replaced by a spider-fine thread of tungsten wire). Carbon, when overheated with too much voltage, evaporates. It condenses, as a soot, on the inside of the bulb. Edison noticed that there was a clear streak in the carbon deposit in line with the edge of the loop of the filament. The carbon atoms were being thrown off from one side of the loop, and the other side of the loop was casting a "white shadow." Edison fixed a metal plate between the legs of the carbon "hairpin" that was serving as his filament. To this plate he fixed a platinum wire. He found that when the filament was made to glow by direct current (flowing only in one direction) a sensitive instrument connected with the plate and also with the "inlet" terminal of the filament indicated a strong current, but that if the plate and the "outlet" terminal were connected there was no current.

Edison did not explain it or pursue it, but, after Thomson had established the existence of electrons, the British physicist Sir Ambrose Fleming recognized the meaning of the "Edison effect." He placed a metal cylinder around the filament and found that negative electricity would pass from the filament to the cylinder but not the other way. Fleming called the device a diode because it contained two electrodes, the cylinder and the filament. He also noted that when an alternating current was applied, only the positive halves of the waves were passed—that is, the wave was rectified. Lee De Forest, a U.S. electrical engineer, improved Fleming's "diode" into a "triode." He introduced a "grid" that made possible not only rectification but also amplification of the signal. This thermionic (heated-wire) valve device revolutionized communications and reigned for fifty years, with more and more refinements.

Thermionic valves require large glass vacuum tubes, and these generated heat that had to be removed. During World War II, when airborne radar and proximity fuses required miniaturization, the tubes of toughened glass were reduced to finger-joint sizes.

Transistors

The triumph of miniaturization came after the war with the development of the transistor. One of the great inventions in the history of electronics and electrical communications, the transistor outranks in importance even the thermionic vacuum tube, which first made possible radio and television. The inventors of the transistor—the Americans John Bardeen, Walter H. Brattain, and William B. Shockley—received the Nobel Prize for physics in 1956.

The transistor immediately became the main basis of all electronics. Not only did it supplant the tube in nearly every application but its particular advantages also made possible a great extension of the scope of electronics. Formerly, elec-

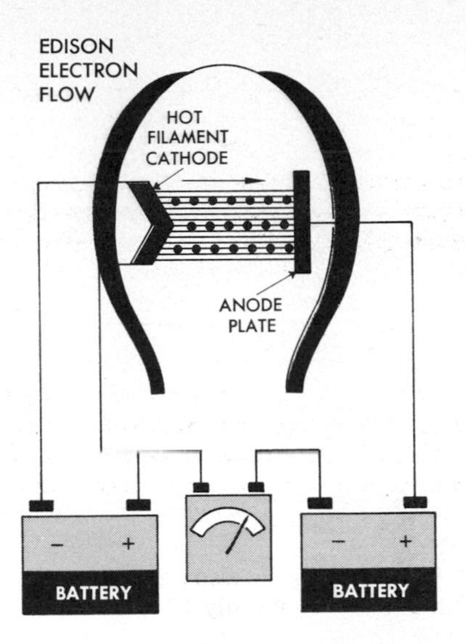

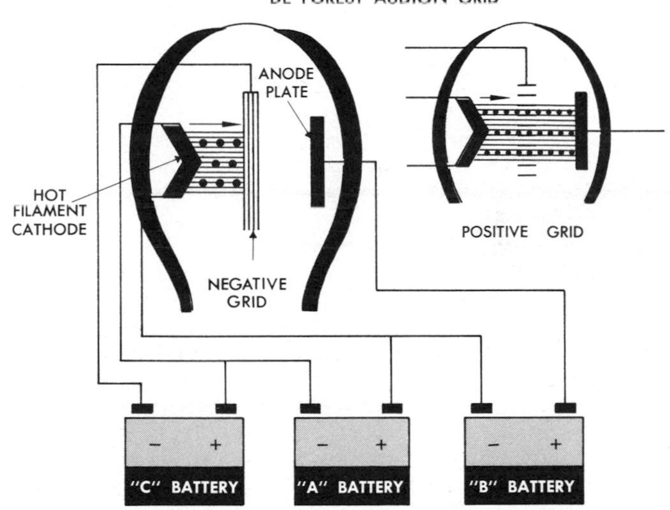

The early Edison electronic tube was a diode (top). When voltage was applied, electrons flowed from the negative hot filament cathode to the positive anode, or collector, plate. In DeForest's triode (bottom) the third electrode was a grid, or fine wire mesh; a negatively charged grid stopped the flow of electrons while a positively charged grid released it.

Silicon wafers are placed in an ion implantation machine during fabrication of large-scaled integrated circuits used in microprocessor memories. The development of the microprocessor is the result of a series of technological advances that began with the introduction of the transistor and culminated in large-scale integrated circuits.

tronics had dealt mostly with the transmission of messages by means such as telephone, radio, television, and telegraph. With the transistor, the scope of electronics broadened to include computers and logic, switching, control, and all kinds of automatic and goal-seeking mechanisms. Most of the new applications became feasible only because of the availability of transistors and, later, integrated circuits.

The invention of the transistor in December 1947 resulted from a program of theoretical and experimental investigation into the properties of semiconductors at the Bell Telephone Laboratories. So called because they are intermediate in electrical conductivity between metals and insulators, semiconductors have a long history of use in electronics. Point-contact rectifiers ("cat-whiskers") were used in experimental detection of Hertzian radio waves as early as 1904. Semiconductors were used as radio frequency detectors and

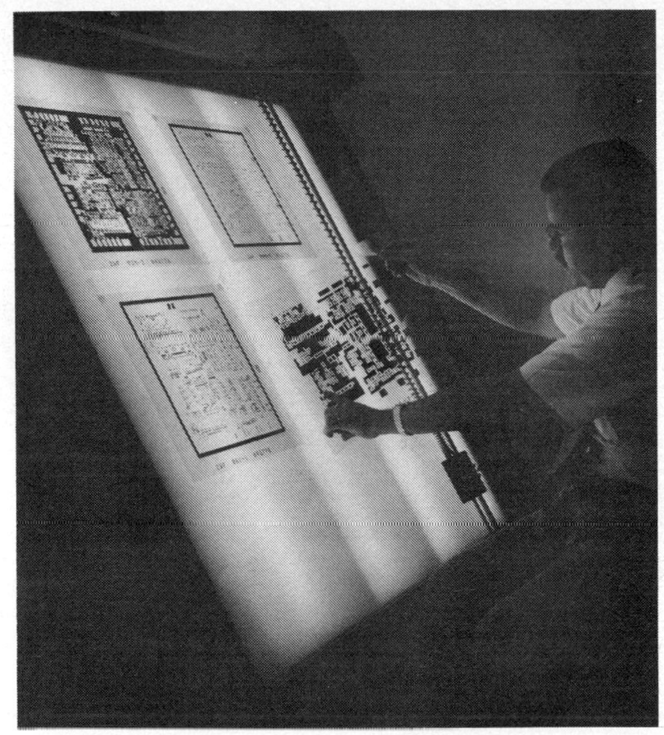

Plastic sheets with microcomputer circuits imprinted in various colors are used to check the accuracy of a process in which the circuits are photoengraved on a silicon chip.

also commercially as power rectifiers (made of selenium) and telephone rectifiers (made of copper oxide or silicon carbide). Semiconductors became increasingly important in communications during World War II when it was found that crystal detectors were much more sensitive for the detection of radar microwaves then vacuum tubes.

In a typical vacuum tube a stream of electrons released from a source (cathode) is controlled by a small input signal applied between the cathode and a control grid. These controlled electrons are then accelerated and collected by a positively charged electrode (anode). The energy gained from the anode appears as a larger power facsimile of the input signal in the output load. Transistors, on the other hand, achieve their amplification functions by controlling the flow of electrons within the specially treated solid materials called semiconductors. In a typical transistor the charge carriers are released from a source (emitter) and controlled by a signal between the emitter and a control electrode (base). The controlled stream is then accelerated and collected by a relatively high-potential electrode (collector), and, again, the energy gained from the collector appears as an amplified replica of the input signal in the output load.

The operation of typical transistors depends on the fact that semiconductors such as germanium or silicon can conduct electricity by the motion of either or both of two kinds of charge carriers. One kind, which carries a negative charge, is called a conduction electron. This is an electron of such an energy that it can wander through the crystal with relative freedom. The other kind, which carries a positive charge, is called a hole because it is actually a deficiency of one electron in the valence bond that ties the crystal together. The motion leaves another deficiency behind; thus, in effect, the hole moves through the crystal carrying a positive charge. If the appropriate chemical impurities or other deviations from lattice perfection are incorporated into the crystal, either type of charge carrier can be caused to predominate, with the other type being in the minority. Thus according to whether the negative or positive carriers are in the majority, a specimen of semiconductivity is classified as N-type or P-type. Most transistors are constructed by juxtaposing regions of different conductivity types within a single crystal of semiconductor so as to produce the useful controlled current of charge carriers.

A simple example of the many types of transistors is the PN junction diode, which is not only a useful device in itself but is also a building block with which to construct other semiconductor devices. When suitable regions of P- and N-type conductivity adjoin in the same semiconductor, the magnitude of current flow depends markedly on the direction of the potential applied. For instance, if the P region is made positive, then some of the positive charges (holes) it contains are impelled across the PN boundary into the negatively polarized N region. Concurrently, some negatively charged conduction electrons from the N region are attracted across the boundary into the P region. In general, flows of both holes and electrons may contribute appreciably to the electric current that flows in this direction of applied potential, called forward bias. On the other hand, if the applied potential is reversed, then both types of majority carriers—holes in the P side and electrons in the N side—are repelled from the boundary. Only the minority carriers—electrons in the P side and holes in the N side—are attracted across the boundary. Since normally these minority carriers are very small in number, the junction draws comparatively little current under reverse bias conditions.

Because of these properties, the PN junction diode is widely used as a rectifier. In ordinary radio receivers, for example, rectification is used in developing direct power supply voltage from the alternating-current supply mains. Other rectifiers (detectors) recover the sound signal from an amplitude-modulated (AM) radio broadcast wave, while still others (mixers) are used to shift the frequency of the radio wave to a value desired for amplification purposes.

The PN junction is also an important feature of the action of many transistors in that the junction can act either as a source or as a collector of charge carriers. Under forward bias the current that flows consists of the injection or emission of minority carriers, both holes into the N region and electrons into the P region. The concentration of minority carriers thus can be greatly increased over that normally present. Moreover, if the semiconductor is a nearly perfect single crystal, the injected minority carriers may remain above the normal concentration an appreciable length of time; lifetimes of 0.1 to 100 microseconds (μsec) are commonly used in transistor material. If too long a time elapses, however, the minority carriers will each recombine with a majority carrier and tend to reduce the minority carrier concentration toward its equilibrium value.

Similarly, the PN junction under reverse bias acts as a collector of minority carriers since any that may arrive at the junction are attracted across it. Under normal diode conditions this reverse current of collected minority carriers is low, reflecting the low equilibrium concentration of such carriers. If the minority carrier concentration is increased through the action of a nearby emitter, however, the collector current increases correspondingly.

Sensors

Since physical influences other than direct or alternating voltages also can affect semiconductors, there exist a number of transistor-related devices that can be used as sensors for light, mechanical stress, magnetism, and nuclear radiation. A great variety of machines known as automatons, or goal-seeking mechanisms, can be built by combining one or more sensors, a data processor to combine the signals from the sensors with other data as desired, and an actuator to drive some external mechanism as desired from the result. Such goal-seeking mechanisms are of increasing importance both technically and sociologically.

Light-Sensing Devices. Light of sufficiently high frequency is converted directly into electrical output. The light can excite a valence electron up into the conduction band of a semiconductor, thus producing a hole-electron pair consisting of the new conduction electron (negatively charged) and the positively charged hole in the valence band representing the missing valence electron. Both electron and hole can then move in the same way as in a transistor; consequently, transistors and diodes are sensitive to light of sufficiently short wavelength. In fact, some photodiodes can serve as secondary standards for the measurement of light flux, since their response is accurately one hole-electron pair produced for each quantum $h\nu$ of light actually absorbed (h = Planck's constant and ν = frequency).

The solar cells that converted sunlight into electrical power in the first *Telstar* microwave communications satellite consisted of wafers of P-type silicon with a thin N diffusion at the surface. Full sunlight will produce about 50 milliamperes at 0.5 volts from such a cell with an area of about 2 square centimeters. Connection of a number of these cells in series makes a solar battery. The Telstar batteries produced about 12 watts of primary power to operate the communications equipment.

Electroluminescence. The reverse effect also occurs. Certain diodes emit light when energized by injecting electrons and holes into the same region. This causes radiative recombination to occur; that is, the conduction electron drops into the hole in the valence band, liberating a quantum of light energy.

Because nonradiative recombination can also occur, at room temperature only a small fraction—typically 10^{-5} for gallium phosphide—of the recombining electrons emits light. Nevertheless, the resultant signal is easily visible and usable for compact signal lamps, for instance. At lower temperatures, where there is less thermal agitation in the crystal, the nonradiative recombination is greatly reduced, and the light-

The successful launching of Telstar *inaugurated a new age in the field of electronic communications. The satellite was powered by nickel-cadmium batteries, was recharged by 3,600 solar cells, and contained more than 1,000 transistors.*

emitting efficiency may approach 100% at temperatures of liquid nitrogen and below.

Injection Lasers. When the luminescent efficiency is high enough, laser (light amplification by stimulated emission of radiation) action may occur. As in other lasers there must be a population of excited electrons ready to emit light. In the injection lasers the injected holes and electrons, ready to recombine, fulfill this need. The carriers are injected into the semiconductor by forward-biasing a semiconductor junction, just like the carriers injected into a transistor.

Silicon and germanium, the best materials for transistors, however, are not good for lasers because their recombination between holes and electrons is indirect; that is, the carrier momentum of the lattice atoms is enlisted to equate the momenta of electron and hole so that they can recombine. A direct material such as gallium arsenide in which such assistance is not required for hole-electron recombination is needed for laser action. Operation is quite efficient at liquid nitrogen temperatures (−320° F, or −195° C) but is difficult to attain at room temperature because of the rapid increase of nonradiative recombination with temperature. Room-temperature operation may be obtained in short pulses by using high current drive. In 1970 continuous room-temperature operation of an injection laser was first achieved by using an elaborate structure of several layers, with gallium arsenide sandwiched by gallium aluminum arsenide.

Photoluminescence. Except that the active plasma of holes and electrons is generated by light instead of by electrical current, photoluminescence is like electroluminescence. In a ruby laser, for example, intense illumination by xenon flash lamps can build up an active plasma. When the laser is switched on, this energy is all discharged at once in a huge burst of red light, the momentary intensity of which may reach one million kilowatts.

Ruby is used because it contains ions (chromium) that can store the excited electrons in their inner atomic shells. Such lasers have many industrial applications, including such odd ones as drilling small holes through diamonds.

Magnetic Sensors. A current flowing across a magnetic field experiences a transverse deflecting force. In a semiconductor device this transverse deflection causes a transverse electric field to appear, and in this way the device is said to sense the presence of the magnetic field.

Such devices can be used to measure magnetic flux. They are compact, convenient, and have no moving parts. Their signals also can be amplified and used to trigger large amounts of power.

Electron Sensors. A beam of electrons in a vacuum may fall upon a semiconductor and produce a variety of transistor-related effects. The charge so injected, for instance, can be collected by a reverse-biased junction in full analogy to the charge injected by an emitter into a transistor or into a photocell by a light beam. A particularly interesting example is the vidicon-type camera tube. In this tube the image to be televised is focused on a target that consists of a thin plate of silicon less than two centimeters square, a little smaller than a postage stamp. Into the N-type material have been diffused small P areas, a total number of about 800,000 diodes, each of which will be one dot in the eventual television picture. At each dot the light from the corresponding portion of the image liberates a number of electrons. Periodically, a finely focused electron beam scans the target once each 1/30 of a second. Wherever the beam hits a diode it brings its potential back to a standard value (the voltage of the cathode from which the beam was emitted). During the next 1/30 second the potential climbs back up to a value that depends on the light; therefore, to reduce the diode voltage back to zero the beam must supply to each dot a current pulse proportional to the light intensity. As the beam sweeps along, the varying signal supplied to the dots is picked up by a contact on the edge of the target. This is the picture or "video" signal that can then be amplified and transmitted in the usual manner. Compared to other vidicon tubes, this one, whose target uses transistor-type techniques and materials, has the advantage of being made of refractory, high-temperature materials; they allow the tube to be pumped at high temperature to give a "hard" vacuum and corresponding long life.

The familiar picture tube of television receivers reproduces the picture by electroluminescence. As the high voltage beam scans across the face of the tube at a potential of about 15,000 volts or more, each electron excites in the semiconductor phosphor a large number of secondary electron-hole pairs—one for every 5–10 volts of energy, depending on the properties of the target material. A small fraction of these excited pairs recombine radiatively, producing enough light for the television viewer to see the picture.

Photoconductors and Xerography. Materials that lose electrical resistance when subject to illumination are photoconductors. The phenomenon is most striking when the unilluminated material has low conductance, since the mobile carriers created through absorption of light quanta can then produce a drastic reduction in resistance. A good photoconductor is usually an intrinsic semiconductor; each photon (light quantum) absorbed by the material excites an electron across the energy gap to create an electron-hole pair.

One of the interesting applications is that of xerography, a photocopy process. If the photoconductor has a large enough resistance in the dark, it can store a charge for appreciable periods of time. If a thin layer of photoconducting material is charged electrostatically and then illuminated with a focused or projected image, the bright regions become electrically conducting and discharge. After the illumination is removed the photoconductor can be dusted with a frictionally charged powder that adheres to the remaining surface charge. The photoconductor thus carries a powder reproduction of the original image, which can be transferred by contact to paper and fixed by heat to provide an inexpensive photocopy. Since some of the powder tends to remain on the paper, a number of copies of the same image can be made.

Chemical Electricity

A fuel cell is a primary electrochemical cell in which electricity is produced directly by the reaction of a gas or liquid fuel supplied to one electrode and oxygen or air supplied to the other. To be continuously useful, the electrodes and the electrolyte between them should be unchanged by the reaction. In this respect the fuel cell differs from other primary cells in which active ingredients are incorporated within the electrodes and are chemically changed during the reaction. When they are depleted, the cell must be replaced. In secondary cells, the electrodes and electrolyte can be restored by charging, that is, by passing a current in the direction opposite to that passing during discharge.

The first battery of fuel cells, built in 1839, produced electric current from hydrogen and oxygen supplied to inert electrodes in sulfuric acid. During the nineteenth century there was a continuing effort to produce electricity from direct electrochemical oxidation of a conventional fuel such as coal or from the carbon monoxide and hydrogen derived from coal. Molten salts as well as aqueous solutions were tried as electrolytes.

Although all the important principles of operation were recognized and great ingenuity was applied to development, none of this early work led to devices able to compete with generators driven by steam or waterpower. Reactions were inefficient, rates were too slow, and cell life was too short.

After World War II there was a great increase in interest in the development of practical fuel cells and fuel batteries. As early as 1932, researchers at Cambridge, England, had begun to investigate the possibility of building practical fuel batteries capable of supplying large power. Their success with high-temperature, high-pressure cells stimulated other work in the U.S.S.R., West Germany, and The Netherlands, aimed toward the large-scale production of electricity. Beginning in the late 1950s, the promise of improvements resulting from new technology, the demanding requirements of military and space vehicle programs, and the drive to reduce atmospheric pollution from power plants and internal combustion engines resulted in the expenditure of hundreds of millions of dollars for fuel cell research and development. Many fuel cells were shown to be operable, and a number of them were reduced to practicable systems. Hydrogen, carbon monoxide, methyl alcohol, hydrazine, and some of the simpler hydrocarbons were used directly as fuels. In so-called indirect cells, conventional fuels or other materials, such as ammonia, were treated chemically to provide hydrogen, which was then used in the fuel cell. Air and oxygen were used as oxidants. A variety of electrolytes were employed, including concentrated alkaline or acid solutions usually at temperatures below 300° F (148° C), molten carbonates, and other fused salts at temperatures of several hundred degrees Celsius. Certain cells used ionically conducting modified zirconium oxide as a solid electrolyte at temperatures near 1,800° F (982° C). Electrodes usually consist of porous metal or carbon and at the lower temperatures include catalysts to increase reaction rates to reasonable levels.

A limited number of fuel cell types were built into complete operating battery systems, incorporating means for the storage and the controlled supply of fuel and oxidant and for the removal of heat and reaction products. Two such systems, using liquid hydrogen and oxygen, successfully supplied electrical power for U.S. manned space vehicles. The *Gemini* spacecraft employed low-temperature cells in which the electrolyte was an ion-exchange membrane. The electrodes were of platinum black and polytetrafluoroethylene. The larger *Apollo* spacecraft fuel battery used a system based on porous nickel electrodes and a potassium hydroxide electrolyte operating at 480° F (248° C) and higher. Smaller systems of 60–240 watts, developed to power portable military radio sets, used either hydrazine or hydrogen derived from metal hydrides as fuel, with air as the oxidant.

Most of the technological problems of practical fuel batteries apparently can be solved, especially for fuels such as hydrogen, hydrazine, and methyl alcohol. Cheap fuels, however, seem to require expensive catalysts and high-temperature operation. Major engineering breakthroughs are required, therefore, before fuel batteries can compete with the internal combustion engine as a versatile power plant for the automobile.

9.
The First Atomic Explosion—A Benchmark of History

The first atomic explosion and the development of atomic weapons; attempts to forge an effective atomic weapons treaty; and the peaceful use of atomic energy, including the problem of waste disposal

Precisely on the last countdown second, at 5:30 A.M. on Monday, July 16, 1945, energy was released from the nucleus of the atom. Never has the beginning of a new era been so accurately dated. That explosion, on the Alamogordo Air Force Base 120 miles south of Albuquerque, N. Mex., was a benchmark in history from which the future of mankind would thereafter be dated.

Life on the Earth had suddenly and irreversibly changed. Man had acquired a new source of energy and with it the power to veto the evolution of his own species and to turn his planet into a radioactive wilderness.

The Atomic Age

In the instant of that first atomic explosion there was an intense white light. The U.S. physicist J. Robert Oppenheimer, who had masterminded the making of that bomb, was clinging to a post 3½ miles away waiting for the shock wave. He found himself recalling the Bhagavadgītā, the Hindu Song of God: "And if the brightness of a thousand suns were burst at once upon the sky, that would be the vision of the Mighty One," and in the reverberations of the tremendous roar he remembered another quote: "I am become as Death, the shatterer of worlds."

After the explosion a ball of fire rose rapidly, followed by a mushroom cloud extending 8 miles up into the stratosphere. The steel tower on which the device had been mounted was completely vaporized, and the surface of the surrounding desert was fused to glass for a radius of about 800 yards.

This was a secret test. The news was broken to the world at large on August 6, when a bomb the equivalent of 20,000 tons of TNT, the most powerful chemical explosive, was dropped on the Japanese city of Hiroshima. It devastated four square miles of the heart of the city. Of the population of about 343,000, approximately 66,000 were killed and 69,000 were injured. More than 67% of the city's structures were destroyed or severely damaged. On August 9 a second bomb was dropped on Nagasaki, where approximately 39,000 persons were killed and 25,000 were injured. About 40% of the property was wrecked. To the devastation was added another dimension—radiation, affecting not only those exposed but capable of producing mutations in future generations.

The Atomic Age was ushered in. There had been no preparation, no forewarning, because the Manhattan Project that produced the bomb had been conducted behind the sky-high walls of military secrecy. Studies carried out by the World Health Organization (WHO) of the United Nations showed that not only in the advanced countries but also among simpler peoples throughout the world the effects had been profound. To many people one of the most alarming aspects of the atomic bomb was the radiation, unseen, unfelt, unsmelled, and untasted, but all-pervading. The WHO study group on the mental aspects of atomic energy commented: "We were back in the childhood of mankind, in the dark caves of our own emotions." Our primitive ancestors had feared and tried to appease the elemental gods of fire and flood but the man-made modern elements were radioactive. This "nuclear superstition" still persists because scientists and authorities have not in the decades that have passed been able to convince people that the fears are unfounded, to rebut superstition with rational assurances. This has led to resistance to the siting of nuclear reactors for industrial purposes, doubts about new types of reactors, and concern about the movement and disposal of radioactive wastes from the fission processes. Nor, of course, can one dismiss the risks of the existence of tens of thousands of strategical and tactical nuclear weapons ready for military use and deployed throughout the world.

After the war the United States continued its testing of nuclear fission weapons on Pacific islands and in Nevada. The United Kingdom, the Soviet Union, France, China, and India produced and tested bombs.

In the early 1950s the public became aware of another type of bomb, variously known as the hydrogen or H-bomb, the superbomb, the fusion bomb, or the thermonuclear bomb. On Nov. 1, 1952, the United States carried out an H-bomb test in the Pacific with an estimated equivalent of 5 to 7 million tons of TNT. The explosion obliterated an island and left a hole more than fifty yards deep and almost a mile in diameter.

On March 1, 1954, the United States tested a fusion bomb of 12 to 14 megatons, twice the previous explosive power. In spite of the official assumptions that the fallout (the radioactive debris of the bomb) would be confined to the immediate locality, radioactive ash covered a Japanese fishing boat about sixty miles distant. Twenty-three fishermen were seriously affected, and one of them died. The official assurances, apart from the localization of the fission fallout, had been that the bomb would punch a hole in the stratosphere where any radioactive gases would dissipate and would not reenter the atmosphere. This was a miscalculation on two counts: radioactive krypton gas from the bomb decayed into radioactive strontium, and the radioactive strontium returned to the atmosphere and was spread by the climatic jet streams to be deposited as rain throughout the world.

Attempts had been made to minimize the dangers of strontium, which had already been discovered in burros exposed on the test grounds in Nevada, but it now became a major issue. Strontium is an analog of calcium; that is, in the absence of calcium it can take its place. It can help form bones. Therefore, its radioactive version is a "bone seeker" and once in a living body cannot be leached out or otherwise removed. It has an active half-life of twenty-seven years. As a result of this nuclear bomb test and others, radioactive strontium was spread everywhere to be ingested through milk or plants contaminated by the fallout. It was so universal that it has been said that every child whose bones were forming in the years of the frequent nuclear bomb tests in the atmosphere has vestigial radioactive strontium in his or her bones—not necessarily of clinical concern but the brand mark of the Atomic Age.

The Soviet Union tested a fusion bomb on Aug. 12, 1953, and Great Britain tested its first at Christmas Island in the Pacific on May 17, 1957. Three major powers were thus in the possession of weapons capable of vast destructive power.

Nuclear Weapons Treaties

The indiscriminate hazards of fallout from the tests finally convinced the United States, Great Britain, and the Soviet Union of the need for a treaty banning testing in the atmosphere or the sea. They adopted a test-ban treaty in 1963, but it was not accepted by France, and China remained aloof. Both developed and tested thermonuclear weapons in the atmosphere.

It soon became clear that many countries had the knowledge and capacity to produce nuclear bombs. In 1968 a treaty banning the proliferation of nuclear weapons was opened for signature. It provided that nonnuclear powers should renounce any intention of providing themselves with nuclear weapons and agree to submit to control and inspection. The treaty was accepted in principle by most countries, but France and China abstained, and India's answer was to explode its own nuclear weapon. The treaty had the drawback of putting restraints on nonnuclear powers but none on the nuclear powers, not even the obligation not to supply nuclear weapons. The supply of nuclear fuels remained the monopoly of the nuclear powers. It seemed to many a flawed treaty that had no provisions to restrain the arms race between the nuclear powers.

To curb this race the two superpowers, the United States and the Soviet Union, engaged in the long-drawn-out SALT (Strategic Arms Limitation Treaty) talks, which sought to establish a balance of weapons and their delivery systems. These systems by then included strategic bombers; intermediate-range ballistic missiles; intercontinental ballistic missiles; nuclear-powered submarine fleets with nuclear weapons; multiple warhead systems, which from a single carrier rocket can scatter and "home" on ten different enemy targets; and cruise missiles, which under remote or computer-programmed control can elude warning systems and weave their way to a target.

Peaceful Uses of Nuclear Energy

In 1955 the United Nations, frustrated in its earlier attempts to secure international control and inspection of atomic energy, called a conference in Geneva, Switzerland, on the peaceful uses of atomic energy. In general, it was a euphoric occasion. Military secrecy had previously prevented scientists from gathering, but now they could share their factual knowledge. Immediately it became clear that many of the secrets were not secrets at all and that much of the information about the nature of the atom's nucleus was common property arrived at independently. There was a flurry of "declassification" as one scientific disclosure after another made secrecy superfluous.

The overriding concern of the conference was how nuclear energy could be used for human betterment. There was talk about the less-developed countries "making a leap across the centuries" to industrial prosperity with the help of the energy that nuclear reactors could provide. There was talk of "packaging" and setting up reactors in remote places in the deserts and the Arctic, eliminating the need for long-haul road and rail facilities to transport conventional fuels. There was heady talk of using nuclear explosives to turn rivers that flow wastefully to the Arctic toward the steppes and deserts of Soviet Asia. With nuclear power anything seemed possible.

The chairman of the conference, Homi J. Bhabha of India, stated that if thermonuclear fusion could be controlled there would be as much energy as there is deuterium (heavy hydrogen) in the seven seas. That is a lot of energy because deuterium occurs as one part in every 1,600 in water. The conference was taken aback by the pronouncement, but the United States, the Soviet Union, and Great Britain soon coyly admitted that indeed they had research in progress to control thermonuclear fusion. This was to remain a hope deferred for a long time as was the prospect of "packaged" reactors. The packaged reactors indeed materialized but as the power units of nuclear submarines or of surface vessels like the U.S.S. *Savannah* or the Soviet icebreaker *Lenin* and not as the small or moderately rated power stations with which a less developed country could begin to grow industrially.

By the second conference in 1958 the difficulty of achieving controlled nuclear fusion had become apparent. Also the problem of disposing of the radioactive waste products of nuclear fission had become a dominant concern. Thus, at this meeting the euphoria of the first conference had evaporated.

Meanwhile, the process by which the fissile uranium-235 was culled from the much more plentiful uranium-238 and employed to produce the fissile plutonium-239 was being adapted to produce electric power. At an experimental reactor station in Arco, Idaho, electricity from nuclear energy was first supplied in December 1951. In June 1954 a plant capable of supplying the electricity needs of a small town became operable in the Soviet Union. In October 1956 Great Britain's large-scale reactor at Calder Hall in Cumberland County began feeding electricity on a commercial scale into the national grid.

The natural-uranium reactor produces heat as well as plutonium. The reaction is kept under control by means of

The U.S.S. Nautilus, *the world's first nuclear-powered submarine, traveled more than 60,000 miles on its original atomic core, which consisted of approximately eight pounds of uranium. A diesel-powered submarine would have used about 2 million gallons of fuel to travel the same distance.*

cadmium rods that act as "neutron gluttons," gobbling up any excess of neutrons when thrust into the heart of the reactor. The heat is transferred by water or by a molten metal, such as liquid sodium, to heat exchangers. There conventional engineering takes over, and the turbine generators tame the atom into electricity for the home or the factory.

The types of reactors multiplied and were improved. Natural uranium was replaced by enriched uranium, with a high proportion of the fissile uranium-235. Among the various types were pressurized water, sodium graphite, heavy water, boiling water, and fast breeder.

The fast breeder is what had been called in the nursery days of reactors "the pixilated pile" because, like pixies filling the coal scuttle every time the fire is stoked, the breeder reactor produces as much fissile material as it consumes. It is a fast reactor because enriched fuel is used as a furnace. The core is surrounded with a blanket of susceptible material. If urani-

With a 47 million kilowatt capacity, the United States was by far the most important user of nuclear-generated electricity during 1977. Recently, however, the construction of new facilities has slowed down considerably, partially as a result of the government's energy policy.

um-238 is used as the blanket, plutonium is produced; if thorium is used, it is converted into uranium-233, another fissile material.

Thorium is three times more abundant in nature than uranium, and, consequently, if it could be converted to fission fuel the concern about impending shortages might be mitigated. Indeed, that was one of the arguments for breeder reactors in earlier times, but the breeders that were developed were mostly those of the uranium-into-plutonium type. The main reason for this was that plutonium was the preferred product for weaponry.

Radioactive Wastes

The trouble with fission reactors is that after their energy has generated electricity much energy remains as radioactive wastes. In the recycling of the fuel elements a large and dangerous variety of radioactive elements remain. High-level

Continued proliferation of nuclear fission reactors increases the health-threatening hazards of radioactive waste disposal.

radiation wastes are a persisting danger. (Plutonium has a half-life of 24,360 years.) They can be kept liquefied in stainless steel caldrons, but the energetic atoms generate heat and so a continual cooling system has to be maintained. They also generate gases that would be dangerous if they vented. They can be coffined and drowned at the bottom of the oceans, with the risk that they will escape or leach with serious effects on the marine environment. Proposals have been made to put them in rockets and scatter them in space, but there are objections to depositing litter there too and misgivings about launching failures that would scatter them on the Earth. Hydraulic fracture was another suggestion. This would be achieved by drilling bore holes into shale and using hydraulic pressure to shear the shale horizontally so that the radioactive wastes in liquid cement would be spread like a layer in a sandwich to harden. Salt mines suggest possibilities. The preferred preparation, whatever the ultimate disposal, is to glassify the wastes in ceramics from which the radioactive elements would not leach. Whichever approach, however, it is expensive.

It is ironic that, in terms of nuclear fission, man is dealing with the end products of the Big Bang. About 4.5 billion years ago in a spiral arm near the edge of the Milky Way our solar system began to take shape. The star, the sun, formed, and around it "snowballs" of materials formed, collided, and coalesced until some formed the planet Earth. Some of that material was the radioactive nuclear waste of the Big Bang, and it became embodied in the planet's rocks. Man extracted the fissionable ores from those rocks and used them, but now does not know how to put them back in the ground.

10.
Atomic Fusion— Energy from a "Doughnut"

Atomic fusion as a source of energy; plasma, the fourth kind of matter; experiments with "magnetic bottles" and "doughnuts"; and the potential of thermonuclear reactions as a form of safe energy

Gulliver in his travels met a fellow of the Grand Academy of Lagado:

> *He had been eight years upon a project for extracting sunbeams out of cucumbers; which were to be put in vials, hermetically sealed, and let out to warm the air, in raw inclement summers. He told me he did not doubt, in eight years or more, that he would be able to supply the governor's garden with sun-shine at a reasonable rate; but he complained that his stock was low, and entreated me to give him something as an encouragement to ingenuity, especially since this had been a very dear season for cucumbers....*

That has the contemporary ring of modern budget requests, except that they would relate to "doughnuts" and not to cucumbers. Putting the energy of the sun into a doughnut, a toroidal fusion reactor, is extremely expensive. In the twenty-five years after civilian-oriented nuclear fusion research began, the U.S. government invested $1.3 billion in efforts to harness the energy that had exploded in the hydrogen bombs. The ongoing costs were about $300 million a year. Add to that the intensive and expensive experiments being pursued to the same end in the Soviet Union, the European community (notably in Great Britain and in West Germany), and in Japan, and one can see that Gulliver's small present to buy cucumbers would not go far.

To grasp the relevance of doughnuts it is necessary to recall the processes, already discussed, by which the sun produces the energy that comes to us as sunbeams. This is a slow process, a gavotte of atoms slowly progressing through eons of time. Carbon (^{12}C) links up with a proton (the hydrogen nucleus) and forms an isotope of nitrogen (^{13}N). The nitrogen isotope decays into a carbon isotope (^{13}C) by emitting a positron (positive electron) and a neutrino (a particle without mass but having considerable penetrating power). Carbon-13 captures another hydrogen nucleus, forming ordinary nitrogen (^{14}N). The nitrogen next captures another proton to form an isotope of oxygen (^{15}O); this immediately decays, emitting a positron and a neutrino, and becomes ^{15}N. In the final stage the nitrogen captures a fourth proton and yields helium (^{4}He) and ^{12}C. In this last reaction energy is released.

There is another pressure-cooker process going on in the sun at the same time. Two hydrogen atoms without their electrons combine to form deuterium, emitting a positron and a neutrino. The deuterium captures a proton to form an isotope of helium (^{3}He). Two of these isotopes then combine to form ordinary helium by discarding two excess protons.

But there is a simpler and more significant sum. Two deuterium atoms have a combined mass of 4.0324, while the mass of a helium atom is 4.0038. There is a mass difference of 0.0286. According to Einstein's equation relating mass and energy, the mass lost when the two deuterium atoms combine to form one helium atom turns into energy. Though the difference between 4.0324 and 4.0038 does not look like much, it is, indeed, a mighty lot of energy. Scientists want to do that sum, two and two make four, the fast way. That is where the torus, or doughnut, comes in.

The Fusion Process

In order to understand the problem, it is helpful to recapitulate some basic features of the atom. The most common form of the hydrogen atom, or protium, consists of a single electron revolving around a much heavier nucleus, the proton. The rarer forms, deuterium and tritium, have nuclei that consist of a proton plus a neutron and a proton plus two neutrons, respectively, with one orbiting electron. The positive charge of the nucleus attracts the opposite, negative charge of the electron, and thus binds the electron to the atom. On the other hand, similar electrical charges tend to repel one another. Thus, the binding, or fusion, of two nuclei is difficult because the two positive charges repel each other.

Deuterium and tritium fuse more readily than other combinations of nuclei, and fusion reactor designs are therefore generally based on this process. When a deuterium and a

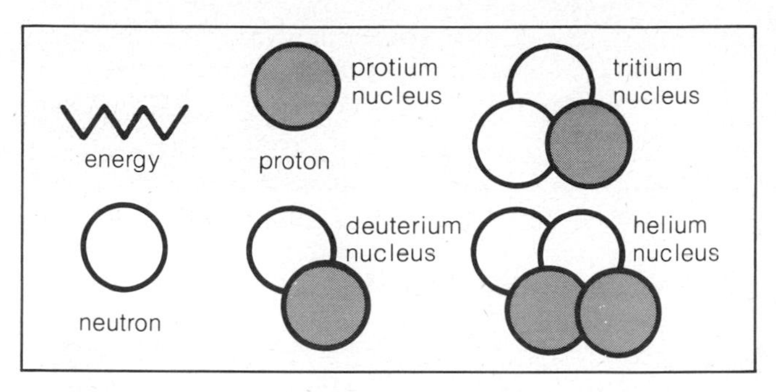

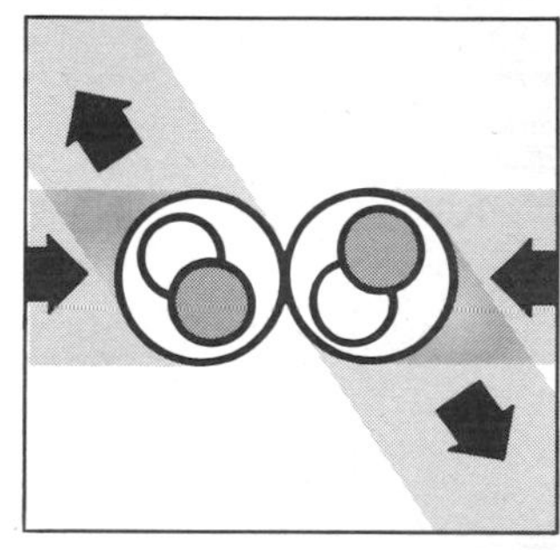

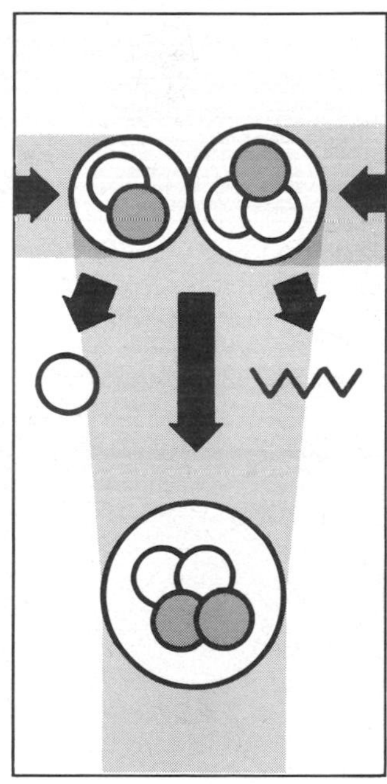

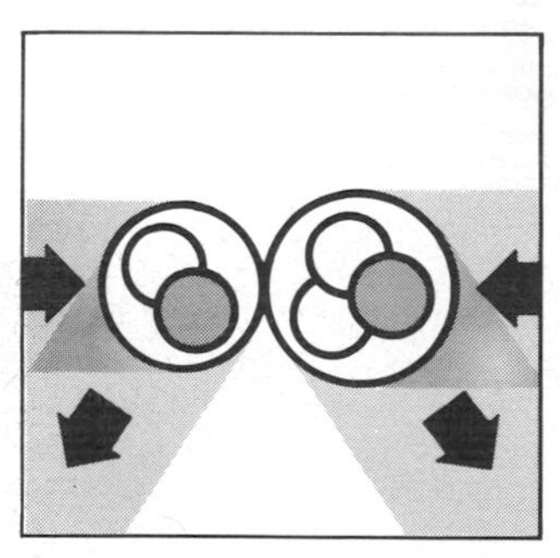

The nuclei of the hydrogen isotopes—protium, deuterium, and tritium—consist, respectively, of a proton, a proton plus one neutron, and a proton plus two neutrons (top). The positively charged nuclei, each orbited by a negatively charged electron, naturally repel each other (left, center and bottom), but if their energies of approach are sufficiently high they fuse—yielding a helium nucleus, a neutron, and vast amounts of energy (lower right).

tritium nucleus meet, however, they are most likely to glance off each other without fusing. Only if their energies of approach are sufficiently high is there any appreciable chance that they will come close enough to experience the nonelectrical forces that give rise to nuclear fusion. The minimum energy of motion required for the deuterium-tritium reaction to take place is about 10,000 electron volts (eV). This is the energy acquired by an electron or proton in a 10,000-volt (V) accelerator, or equivalently, it is the average energy of motion of particles at a temperature of about 100,000,000° C. Heat, as the English philosopher Francis Bacon said nearly four centuries ago, is "motion and nothing else."

The term *thermonuclear reactions* refers to fusion reactions that take place as a result of this thermal motion. In addition to deuterium-tritium, deuterium will also react with itself but only at temperatures above 1,000,000,000° C.

The deuterium-tritium reaction yields a helium nucleus, a neutron, and 17.6 million electron volts (MeV) of energy, most of which is carried off by the neutron. Thus, potentially there is roughly a thousandfold increase over the initial investment of 10,000 eV in the energy of motion of the deuterium and tritium nuclei. The actual energy multiplication factor obtained depends on the fraction of the hot fuel that reacts before escaping or cooling.

It is necessary that the reactor should confine the hot fuel long enough so that a sufficiently large fraction will react and thereby release more energy than was invested in fuel heating. For the deuterium-tritium reaction the fuel confinement time, expressed in seconds times the density of nuclei per cubic centimeter, should exceed 1,000,000,000,000,000 (10^{15}). During the period 1960–70 it became relatively easy to achieve high temperatures and make fusion reactions in a laboratory experiment, but the best numbers of nuclei that were achieved were in the range of 10^{11}–10^{12}.

In order to appreciate the difficulties involved and the nature of the experimental breakthroughs that have to be made, it is necessary to understand some basic properties of high-temperature matter. We are dealing with the fourth state of matter—solid, liquid, gas, and, now, plasma. When a liquid is heated sufficiently, it becomes a gas—the atoms break free of each other. When a gas is heated further, it becomes a plasma—the constituent particles of the atoms, the electrons and nuclei, break free of each other. Free nuclei are also called ions, and the breakup process is called ionization. In hydrogen, ionization occurs at a temperature of about 10,000° C, relatively low when compared to the 100,000,000° C temperatures required for fusion. On the Earth natural plasmas are encountered quite rarely, for example, in lightning strokes and in the northern lights. In the universe as a whole, however, the plasma state is by far the most common condition of matter. The sun and other stars are giant fusion reactors, and the spaces surrounding them are filled with dilute plasmas.

Plasma produced in a fusion reactor on Earth cannot be confined by walls made of ordinary solid matter because the contact with the walls would cool the plasma and convert it back into a gas. Fortunately, there is a natural way to confine the charged electrons and ions of a plasma: by means of a magnetic field. In a uniform magnetic field a charged particle must gyrate locally around a magnetic field line, though it can travel freely along the line. This constraint on the motion of plasma particles across magnetic field lines is the basis for the concept of the "magnetic bottle."

Fusion power can also be produced by freely exploding

A laser fusion research system being built at Los Alamos Scientific Laboratory in New Mexico is designed to utilize inertial confinement and laser heating to achieve thermonuclear reactions. The long-range goal of the program is to develop a nuclear fusion power plant.

plasmas—not confined by a magnetic bottle—provided the initial plasma is sufficiently dense. For example, at the normal density of solid deuterium–tritium (4×10^{22} atoms per cubic centimeter), the reaction time is only several times 10^{-9} seconds at 100,000,000° C. During this time the plasma expands by only a few millimeters. The difficult problem is how to heat even a cubic millimeter of such a plasma to 100,000,000° C in less than 10^{-9} seconds, since this requires a highly focused energy input of approximately 10^5 joules at a heating power of 10^{14} watts. (A joule is a unit of work equal to approximately 0.7375 foot-pounds.) The recent development of high-powered lasers is beginning to bring this achievement within the realm of technological feasibility.

Magnetic Bottles

One way to prevent the plasma particles from escaping is to bend the magnetic field lines in a circle, making a closed, toroidal (doughnut-shaped) magnetic bottle. Another solution is to leave the magnetic bottle open-ended but make the magnetic field lines converge toward each end. This is called a "mirror machine." How the mirror machine contains a plasma particle can be understood by an analogy with a compass needle between the poles of a C-magnet. A gyrating charged particle is a small magnetic dipole, as is a compass needle; furthermore, the particle dipole is always diamagnetic (like a compass needle that is oriented so as to repel the two poles of the C-magnet). Consequently, the gyrating particle stays away from the two ends of the magnetic bottle. This confinement effect, however, does not work for particles that are simply flying straight out along a field line instead of gyrating around it. In this sense the mirror machine is a leakier magnetic bottle than the torus, which will confine a particle no matter in what direction it is moving.

In the thermonuclear reaction process energetic nuclei glance off each other many times by electric repulsion until they happen to make a sufficiently close collision so that they fuse. Unfortunately, the process of bouncing helps particles to escape rather rapidly from a mirror machine. Two gyrating nuclei may collide in just such a way that they both escape straight out the ends. In a toroidal bottle, the collision process gives rise to a more gradual diffusion of particles across the magnetic field lines. In either case scientists have calculated that plasma confinement times will be long enough to make reactor operation possible. This is known as the "classical"

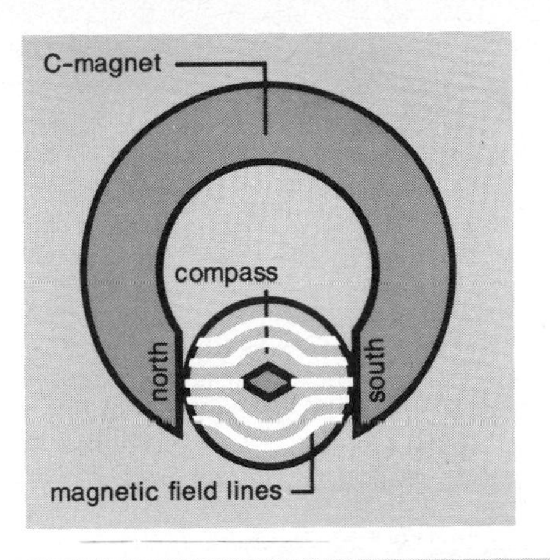

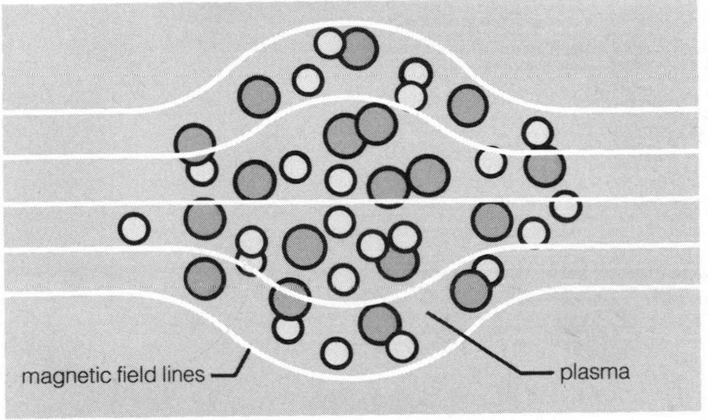

A magnetic bottle of the mirror type has nonclosed magnetic field lines that converge toward each end of the bottle (bottom), corresponding to an increase in field strength. A gyrating charged particle, which resembles a diamagnetic compass needle (top), is repelled by the ends of the bottle as the needle is by the ends of a C-magnet.

confinement prediction for fusion reactors.

The main obstacle to progress with fusion power has been that actual plasmas escape far more rapidly from magnetic bottles than one would have expected from the classical prediction. One of the most important obstacles has been magnetohydrodynamic (MHD) instability. MHD is the term applied to the motions of electrically conducting fluids—liquids, gases, and, here, plasmas. These motions cause plasma to escape rapidly from confinement.

In coping with MHD instabilities an important breakthrough came in 1961 with the announcement of the first "magnetic well" experiment, carried out at the Kurchatov Institute in Moscow. As had been noted for many years by theorists in the United States and the Soviet Union, there exist special kinds of mirror configurations in which the magnetic field strength increases in all directions away from the plasma; because of this increase in strength a plasma confined in such a magnetic well would be expected to be absolutely stable against MHD modes. The Kurchatov experiment was able to operate alternately in the simple mirror machine configuration and in a magnetic well configuration. A dramatic stabilization effect and increase of plasma lifetime was observed in the latter case. The typical plasma values of the PR-5, a more advanced embodiment of the Kurchatov experiment, were: ion temperature, 50,000,000° C; electron temperature, 200,000° C; and density, 10^{10} particles per cubic centimeter. The plasma lifetime was 3 milliseconds in the stable configuration and only 0.1 milliseconds in the unstable.

Following the PR-5 results the buildup of hot-iron plasmas in magnetic wells was pursued vigorously, mainly by injection and trapping of energetic neutral atom beams. The Baseball I experiment at the Lawrence Livermore Laboratory in California produced 100,000,000° C plasmas with densities up to 10^9 particles per cubic centimeter and confinement times of about a second. The somewhat similar Phoenix II experiment at the Culham Laboratory in Great Britain reached densities of nearly 10^{10} particles per cubic centimeter. These and other magnetic-well experiments confirmed the complete removal of MHD instabilities as a limiting factor in mirror confinement. Experimental mirror-machine plasmas, however, were not yet able to realize the ideal process of buildup toward very high, collision-limited densities. Instead, the plasma lifetimes and attainable densities were found to be limited by the onset of microinstabilities.

Microinstabilities in a mirror machine take the form of high-frequency electrical waves. These waves are generated in somewhat the same manner in which light waves are generated in a laser. To set the stage for laser action, one needs to create an abnormal distribution of excited atomic states; these can then relax in unison and drive a giant light pulse.

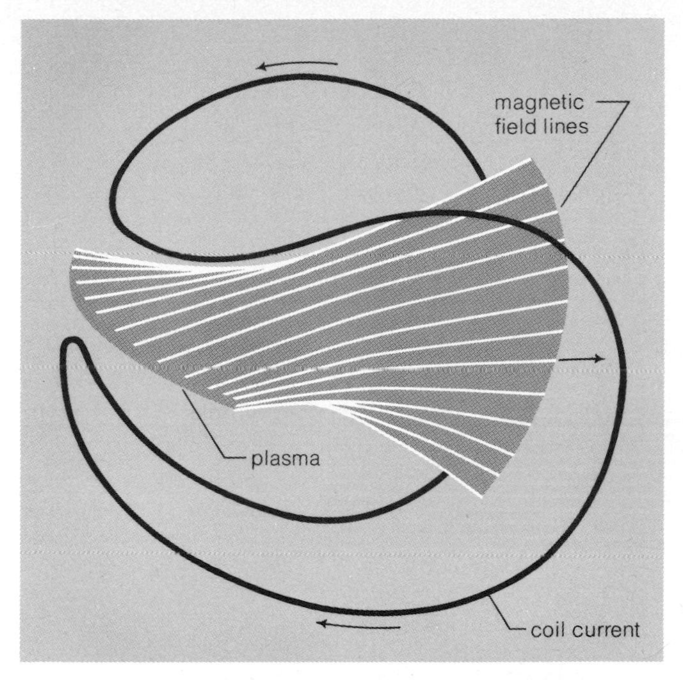

Advanced mirror machine of the magnetic well type has a magnetic field that increases in strength in all directions away from the center of the plasma it confines. The plasma is stable against MHD instabilities.

Because of the leaky ends of the open-ended magnetic bottle its plasma always has an "abnormal" distribution of particle velocities. Large-amplitude plasma waves can arise spontaneously and help to knock plasma particles out of the ends.

The severity of the end-losses depends on how "abnormal" the particle velocity distribution is. The worst distributions are produced by injecting a single beam of energetic particles, all with the same velocity and moving in the same direction. In the best distributions the particles are moving almost randomly in all directions, with a wide range of velocities.

Doughnut Experiments

In a closed magnetic bottle, where plasmas are not lost rapidly by classical scattering, the favored method of buildup toward a hot, dense plasma is by raising the temperature of an initially cold, dense plasma. The plasma, with its free electrons, is an electrical conductor, and so it can be heated by inducing a current to pass through it. Plasma resistance results from the collisions of the current-carrying electrons with stationary ions; the power dissipation associated with the collisions causes heating of the electrons. This is the same process of "resistive" or "ohmic" heating that is employed in a toaster. The heating method is effective up to electron temperatures of about 10,000,000° C, at which point a hydrogen plasma has about as low a resistivity as standard copper. At still higher temperatures the plasma resistivity becomes so low that resistive heating is too slow to be effective.

The magnetic field of a toroidal magnetic bottle must have a rather special structure in order to hold a plasma ring in equilibrium. A simple magnetic field like that surrounding a straight wire is not good enough; with such a field the plasma ring would expand continuously along its major radius (seeking to move in the direction of decreasing magnetic field strength). In order to provide a proper equilibrium the toroidal magnetic field lines must be twisted into the shape of a helix. In the tokamak device, invented in the Soviet Union, this twist is provided by inducing a toroidally directed current around the plasma ring. This current also serves to heat the plasma by the electric toaster method. An alternative approach, invented at the Princeton University Plasma Physics Laboratory, is the stellarator. In this device the helical twist of the magnetic field is caused by currents in helical windings outside the plasma ring. Still other toroidal magnetic bottles use currents flowing in solid rings that are inside a hollow plasma ring, or employ circulating beams of high-energy electrons. The tokamak and stellarator, however, are generally considered the most practical approaches to a first-generation toroidal fusion reactor.

As in mirror machines there is a tendency toward MHD instability. Even worse is the MHD "kink" mode, in which the whole plasma column buckles in the form of a helix. The kink mode occurs when the plasma current of a torus is made too large. Stability against the kink mode is one of the advantages of the tokamak.

At the Kurchatov Institute interesting results were obtained on the T-3 tokamak, a device with a plasma sufficiently large to permit electron temperatures above 10,000,000° C to be reached by ohmic heating when the plasma was at a density of 10^{13}–10^{14} particles per cubic centimeter. Confine-

The Princeton Plasma Physics Laboratory tokamak, a large-torus nuclear fusion device, attained record 60,000,000° C temperatures in 1978.

ment times of about 20 milliseconds were obtained at this record temperature. These remarkable results were repeated and confirmed on the ST tokamak at Princeton.

Larger tokamaks and stellarators were constructed in many countries—including the United States, the Soviet Union, and Great Britain—to take advantage of the improved behavior of high-temperature plasmas. The tokamak, relatively simple and inexpensive, is favored for the initial approach toward developing toroidal reactors. The stellarator, on the other hand, has possible long-range advantages, such as for the confinement of high-pressure plasmas.

Laser Fusion

By the late 1970s the highest densities of hot plasma were being obtained not with pulsed magnetic fields but by heating solid or liquid matter with highly focused external energy sources. High-powered lasers were particularly useful for this purpose. A single neodymium glass laser system could in less than 10^{-9} seconds produce an infrared light pulse of about 1,000 joules focused on a millimeter-long target. When a solid pellet (of about 10^{23} particles per cubic centimeter density) was used as the target, the resultant dense plasma absorbed this incident light pulse with efficiency within a submillimeter surface layer. The resultant plasma had far too high a pressure to be confined by technologically feasible magnetic bottles and thus expanded at thermal velocity.

Fusion-oriented research on laser heating was being carried on vigorously in the United States, the Soviet Union, and France. Neutron yields exceeding 10^5 were reported from deuterium targets. If the target could be precompressed by shock waves to densities higher than those of normal solids, the laser-heating task would be simplified. But even if that were accomplished the technical problems of adapting the laser system to the large-scale production of economical fusion power would remain formidable.

Thermonuclear Engineering

While scientists were wrestling with the basic experimental problems, thought was going into the engineering of industrial fusion reactors. There were many variations in mind. The moderate-pressure toroidal deuterium–tritium reactor can serve to illustrate the kind of engineering problems involved.

The deuterium–tritium reaction produces a neutron of 14.1 MeV energy and an alpha particle (helium-4 ion) of 3.5 MeV. The tritium consumed in a fusion reactor must be regenerated because tritium, unlike deuterium, occurs naturally only in minute amounts. The 14.1 MeV neutron is stopped in a blanket surrounding the plasma volume. This blanket is made up largely of lithium, and in it the incident neutrons breed tritium by nuclear reactions. A blanket of liquid lithium or molten lithium salts is suitable for the breeding purpose as well as for the removal of the heat generated by the energetic neutrons and their secondary reactions. Breeding ratios of 1.3 tritium atoms per incident neutron are readily obtained. The generated heat can be removed from the blanket at a temperature in the range of 500°–1,000° C and used to drive a conventional thermal power plant. The lithium blanket should be about two meters in thickness to prevent any appreciable neutron flux from reaching the magnet coils that surround it.

The first fusion reactor plans, proposed during the early 1950s, had to divert a substantial fraction of the electrical power output to energize the magnetic coil system. The development of superconducting magnets with high field strengths, however, removed this requirement. Magnetic field strengths of 50–150 kilogauss produced by superconducting

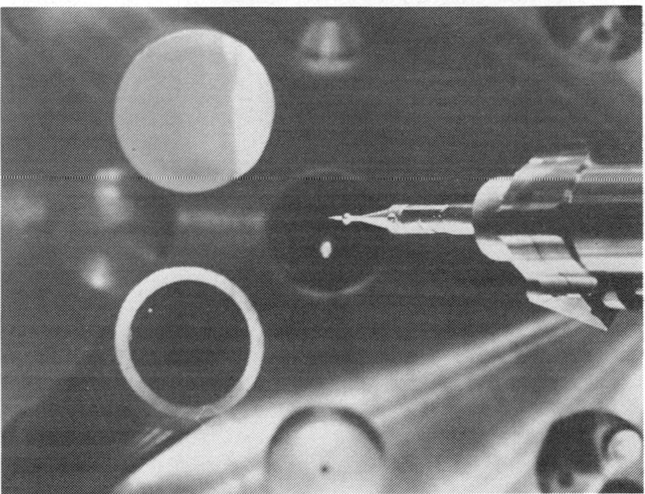

In a high-power laser system built by the Lawrence Livermore Laboratory in California, 20 laser beams (top) are focused onto a tiny hollow glass target that sits on the tip of a positioner (bottom). The laser system's 20–30 trillion watts of optical power compress and heat the fuel inside the target. The resulting fusion releases large amounts of energy.

coils are generally envisaged in present reactor designs. A large fraction of the fusion reactor capital cost is connected with the magnetic coil system and the blanket. In reactors designed to operate most economically the minor radius of the plasma should be at least comparable with the blanket thickness, about 2 yards. The corresponding major radius is 5–10 yards. Such a reactor, given the plasma size, the maximum available magnetic field strength, and the maximum possible ratio of plasma pressure to magnetic field pressure (typically 0.05–0.10, for MHD stability), would provide a total electrical power output in the range of 1–3 kilomegawatts, enough to sustain a population of about two million persons. This is a practical size for modern power stations. The neutron flux through the vacuum chamber surrounding the plasma must not exceed about 10 megawatts per square meter of wall in order to avoid excessive rates of radiation damage.

To start up a fusion reactor a plasma-heating energy investment of about a hundred megajoules is required. Thereafter, the plasma "ignites"; its temperature is maintained or increased further by the 3.5 MeV alpha particles of the deuterium–tritium reaction, which are trapped in the magnetic bottle. The principal process of plasma energy loss is expected to be the diffusion of hot particles. The earliest reactors will probably operate in a slow pulsed mode, burning the fuel as it leaks out and then beginning a new plasma heating cycle. The problem of true steady state operation with continuous fuel injection appears more difficult but can be solved in principle.

The basic fuel, deuterium, can be extracted from water at a price less than a thousandth of the present-day cost of electrical power. The major cost of fusion power will be the capital investment in the reactor, and it is estimated to be competitive with the cost of both conventional and nuclear fission power. The Earth's supply of deuterium will be sufficient for more than 10^{13} years at the present world rate of energy consumption. The known and probable land reserves of lithium would suffice for only about one million years at this rate, but 1,000 times more lithium could be extracted from seawater.

In the discussion of research difficulties magnetohydrodynamics (MHD) has kept cropping up as "instabilities," but it has useful possibilities. At the time when the peaceful uses of fusion energy were first mentioned, Sir George Thomson, a Nobel Prize-winning British physicist, insisted that a thermonuclear reactor could produce electricity directly rather than by removing heat from the reactor and using it to drive turbines. He had in mind MHD.

A gas at a temperature of about 2,500° C is not only incandescent but is also a good conductor of electricity. When such gas is expanded through a nozzle, the kinetic energy of the milling throng of particles is transformed into the directed energy of motion. If the hot conducting gas then moves through a transverse magnetic field, it gives rise to an electric voltage and current, just as can be got from the copper wire in a generator moving through a magnetic field. The ions can be collected by electrodes. Thus the doughnut itself might feed current into the transmission lines.

Fusion and the Environment

Fusion power has inherent economic and environmental advantages that make it a persuasive solution to the world's long-range energy requirements. The basic fuel, deuterium, is inexpensive and abundant. Unlike nuclear fission power, fusion does not produce radioactive wastes except for activated structural materials of the reactor, a problem only at times of maintenance or dismantling. There is no possibility of a runaway fusion reaction. The thermal efficiency of the first fusion reactor designs is comparable to that of conventional power plants, but there is a prospect of future reactor designs with extremely low production of waste heat. The cost of fusion power and the required capital investment are estimated to be competitive with the corresponding figures for present-day conventional power.

The environmental aspects of fusion power are favorable. The amount of fuel present in the plasma is only a few grams, and there is no way in which a nuclear explosion could occur. The only radioactive element or product of the fuel cycle is tritium, which is not a waste product but a valuable fuel to be recycled. The overall biological hazard potential is estimated to be sufficiently low so that in the long run it may be possible to locate fusion reactors close to population centers and utilize any surplus heat.

Remaining environmental defects are all of a kind that can be improved upon by future technology. With advances in fusion reactor design and size, purely deuterium-burning reactors should come into the range of economic usefulness. The problems of lithium consumption, tritium handling, and radioactive waste disposal would then be further reduced.

11.
The Earth as an Engine—Wind, Water, Sun

The historical development of the use of wind, water, and sun as sources of energy; recent developments in harnessing the energy of the Earth; and prospects for the future

The Earth is a solar engine. The energy of the sun drives the winds and generates the waves. It energizes the currents of the oceans as well as the jet streams of the stratosphere. It creates hurricanes and typhoons with pent-up energies immeasurably greater than man-made nuclear bombs. It evaporates water from the oceans to form clouds that will move with weather fronts and precipitate as rain. And in that process it forms thunderheads and releases lightning.

When, therefore, we think of solar energy we should include not just collecting the heat of the sun for domestic or industrial use, or focusing its beams in solar furnaces, but also wind, water, waves, and, with the gravitational intervention of the moon, the tides. In 1961 the United Nations held a conference in Rome on "New Sources of Energy." The irony was that little was said about the new nuclear energy but a great deal was said about the oldest, perennial sources of energy—solar heat, wind, running water, tides, waves, and geothermal energy. Learned experts from the advanced countries told the people from less-developed countries to tap these sources, but as became obvious in the "energy crisis" later, little of this was being done in the industrialized countries.

Power from the Wind

Historically, the use of such sources came surprisingly late in the story of mankind. Surprisingly, because our early ancestors had been ingenious in maximizing muscle power with such simple machines as the lever, the wheel, and the pulley. At the time of the Fifth Egyptian Dynasty, about 2500 B.C., sailing ships were large enough to go to sea. Wind, however, remained for long an auxiliary and not the principal prime mover at sea, probably because oar-pulling galley slaves were cheap.

The earliest known references to windmills are to a Persian millwright in A.D. 644 and to windmills in Seistan, Persia, in A.D. 915. These windmills were of the "horizontal mill" type, with sails radiating from a vertical axis standing in a fixed building that had openings for the inlet and outlet of the wind diametrically opposite to each other. Each mill drove a single pair of stones directly, without the use of gears. The "vertical mill" type, with sails on a horizontal axis, derived directly from the Roman water mill with its right-angle drive to the stones through a single pair of gears. The actual idea of utilizing wind power in this way, however, may have been brought back from the East by Crusaders, and the first European references to such windmills are in France in about 1180 and in England in 1185. The first detailed description from which such a windmill could be built was published in France in 1702, and the first satisfactory working drawings were published in Holland in 1727.

The earliest form of vertical mill was known as the "post mill." It had a boxlike body carrying the sails and containing the gearing, millstones, and machinery. It was mounted on a well-supported wooden post socketed into a horizontal beam on the level of the second floor of the mill body. It could be turned on this so that the sails could be faced into the wind; access to the mill body was by a ladder at the tail of the mill. Usually the post and its substructure were above ground, but sometimes they were buried in a mound. It was then known as a "sunk post mill." The substructure was often protected by a "roundhouse," which was simply a storeroom.

The next development was to place the stones and gearing in a fixed tower. This had a movable top or "cap," which carried the sails and could be turned around on a track or "curb" on top of the tower. Brick and stone towers were usually round; timber towers were usually octagonal and tapering and were known in England as "smock mills." The earliest-known illustration of a tower mill is dated about 1420. Both post and tower mills were to be found throughout Europe and were also built by the British, Dutch, French, and other settlers in America.

At first both post and tower mills drove a single pair of stones for grinding grain through single-stage gearing. By 1430, however, the Low Countries (Belgium, Holland, and Luxembourg) needed more than just animal and manpower for draining the land; the Dutch therefore invented the "hollow post mill." In this design an upright shaft took the drive through the hollow post of the mill and drove a "scoop wheel," resembling a paddle wheel, that scooped the water up from a lower to a higher level. Two pairs of gears were used, at the top and bottom of the upright shaft, respectively, and this opened the way for the larger tower mills with several pairs of stones and ancillary machinery.

Windmills of various designs have been used in many parts of the world to supply power. On the Greek island of Míkonos a stone tower mill is equipped with ten jib sails.

To work efficiently the sails of a windmill must face squarely into the wind, and in the early mills the turning of the post-mill body or the tower-mill cap was done by hand by means of a long "tailpole" stretching down to the ground. Later a series of posts was placed around the mill, a winch fixed to the lower end of the tailpole, and a chain run out from the winch to one of the posts. Winches were also placed in the caps of tower mills and operated from inside or from an endless chain from the ground.

In 1745 Edmund Lee in England invented the automatic "fantail." This consisted of a set of five to eight vanes mounted on the tailpole or the ladder of a post mill at right angles to the sails and connected by gearing to wheels running on a track around the mill. When the wind veered, it struck the sides of the vanes, turned them and, therefore, the track wheels as well; the wheels turned the mill body until the sails were again square into the wind. The fantail was also fitted to the caps of tower mills, driving down to a geared rack on the curb.

The sails of a mill are mounted on an axle or "wind shaft" inclined upward at an angle of from 5° to 15° from the horizontal. On this shaft inside the mill is the first geared wheel, known as the "brake wheel" because a contracting brake acts on its rim. The first mill sails were wooden frames on which sailcloth was spread; each sail was set individually with the mill at rest. The early sails were flat planes inclined at a constant angle to the direction of rotation; later they were built with a twist like that of an airplane propeller.

In 1772 Andrew Meikle of Scotland invented his "spring sail," substituting hinged shutters, like those of a venetian blind, for sailcloths and controlling them by a connecting bar and a spring on each sail. Each spring had to be adjusted individually with the mill at rest according to the power required; the sails were then, within limits, self-regulating.

In 1789 Stephen Hooper in England utilized roller blinds instead of shutters and devised a remote control to enable all the blinds to be adjusted simultaneously while the mill was at work. In 1807 Sir William Cubitt of Great Britain invented his "patent sail," combining Meikle's hinged shutters with Hooper's remote control by chain from the ground via a rod passing through a hole drilled through the wind shaft. The operation was comparable to operating an umbrella; by varying the weights hung on the chain the sails were made self-regulating. Patent sails and fantails, widespread in England, were also adopted in Denmark, Germany, and The Netherlands.

In 1860 R. Catchpole in England successfully applied air brakes to patent sails; the idea was revived in The Netherlands after the application of airfoils to the leading edges of mill sails was initiated by A. J. Dekker in 1923. Others followed and greatly increased the output of mills by enabling them to do useful work in lighter winds. Engineers in England, Germany, and The Netherlands produced sails hinged longitudinally with the same object. Of other varieties of sail the most common are triangular jib sails (eight to twelve are used) wrapped around poles that are braced to a bowsprit extending from the front of the wind shaft. These can be found in Spain, Portugal, the Azores, the Mediterranean islands, Greece, and Turkey.

John Smeaton in England in 1759 was the first to investigate windmill sails scientifically. He preferred five sails, though six- and eight-sailed mills were also built. He and others in England first used cast iron in mill work at this time. At this time, too, the windmill was put to a wide variety of uses by the Dutch in addition to grinding corn and raising water. The most important of these uses were sawing timber, pressing oil from seeds, and paper making. A centrifugal governor was applied to corn mills in England by Thomas Mead in 1787 and was used for maintaining a constant gap between the stones, therefore regulating the fineness of grinding. Wind power was used to drive the sack hoist to raise the grain into bins and so feed the hoppers of the stones by gravity, and automatic bell alarms were provided to give warning when the hoppers needed replenishing.

The annular-sailed wind pump, invented in the United States in the mid-nineteenth century, could lift water to greater heights than the Dutch-type windmill and could operate in low winds.

The annular-sailed wind pump was brought out in the United States by Daniel Hallady in 1854, and its production in steel by Stuart Perry in 1883 led to worldwide adoption.

The design consisted of a number of small vanes set radially in a wheel. A tail vane controlled side-to-side motion, and unwanted twisting motion was governed by setting the wheel off-center with respect to the vertical yaw axis. As the wind increased the mill turned on its vertical axis, reducing the effective area and, therefore, the speed.

Grandpa's Knob. Windmills, however, like the great sailing vessels that once plied the seas, began to lose their dominance as first the steam engine and then gasoline and diesel engines and electric motors took over, supplying the power for most of man's work. Although modern windmills may be built with carefully designed turbine blades and can generate electricity and operate more efficiently than ever before, they have not been able to compete economically with these other engines, which provide power whenever it is needed (the wind not being dependable) and use cheap fossil fuels or hydroelectric energy. By about 1950 interest in wind power had begun to decline drastically, and engineers seeking financial backing for windmill projects found little money available.

The technical feasibility of windmills in certain situations is well established, however. They can be installed in out-of-the-way or even completely isolated locations where quite small amounts of electricity are needed, for example. In such cases they may power meteorological recording devices, radio relay stations, or small navigational aids. It is clear, therefore, that small- and medium-sized windmills may be not only technically but also economically justified under the proper circumstances. What has not been demonstrated is that large wind-powered installations, producing hundreds of thousands or even millions of watts of power, can be operated profitably as part of a major utility network.

The most dramatic attempt to prove the financial value of the large-scale use of wind power occurred in the United States during World War II. A giant wind turbine with two blades forming a propeller measuring 175 feet across (more than half the length of a football field) and weighing 16 tons was built on 2,000-foot Grandpa's Knob, west of Rutland in the Green Mountains of Vermont. The generator became operational in October 1941 and produced power for the Central Vermont Public Service Corporation for sixteen months until a main bearing failed. Wartime material shortages prevented the replacement of the bearing for two years. A month after the machine was repaired one of its blades broke off because of a welding defect, and it was never operated again.

The generator at Grandpa's Knob was built for the S. Morgan Smith Company of York, Pa., and was designed by Palmer C. Putnam; it was therefore known as the Smith-Putnam Wind Turbine Project. Technically it was a remarkable achievement. The best engineering talent of the time, including the world-famous California Institute of Technology aerodynamic expert Theodore von Karman, contributed to the design. After World War II the company undertook an economic analysis of the project and determined that units like that at Grandpa's Knob would cost about 65% more per kilowatt to build than the Vermont utility could afford to pay. Putnam suggested several ways of reducing costs by improving the design, but the Smith Company decided not to invest further capital in the project, which had already cost $1,250,000.

Another economic difficulty with windmills is that, even when they are big, they still do not generate vast amounts of power. The Smith-Putnam turbine, the largest such machine ever built, only produced sufficient electricity "to light a town," in Putnam's words. The output of 1.25 megawatts represented less power than that under the hoods of ten 1979 Cadillac engines run at maximum output. Many large wind generators, therefore, would have to be joined together to make a major contribution to a utility network.

Brace Research Institute at McGill University in Toronto, Canada, began investigating wind-power technology in the early 1960s, particularly in regard to small windmills that could be used in underdeveloped rural areas. They offered plans for "a cheap wind machine for pumping water" whose vertical shaft rotor was made of two drums cut in half and then welded together so that the drum segments formed wind scoops. A "sailwing" windmill utilizing an aerodynamic surface of cloth was developed at Princeton University. It proved to be efficient and also lightweight.

Among the more innovative approaches explored has been one at Montana State University. Engineers there designed airfoils, resembling airplane wings set on end, to be attached to railroad cars. The wind would then move the cars around a five- or ten-mile-long closed track, and the axles of the cars would turn electric generators. The target output of the design was 10–20 megawatts.

A giant wind machine with 89-foot-long blades mounted on a 175-foot-high reinforced concrete tower is currently under construction on the northwestern coast of Jutland, Denmark. At full capacity the huge windmill is expected to produce 4 million kilowatt-hours of electricity per year.

Energy Storage. The conventional way of storing energy produced by a wind generator is in batteries. New combinations of electrodes and electrolytes may improve battery technology, but their future is uncertain.

Water in a reservoir represents stored energy. In the same way air can be compressed for later release. Air could drive a turbine directly, or it might be combined with a small amount of fuel to increase efficiency. At the Hydro-Quebec Institute for Research in Quebec, Canada, researchers have designed a prototype system, driven by a wind machine, to compress air and store it in underground salt caverns. The investigators believed that the system could be enlarged to have a capacity of several megawatts.

Early in this century Danish wind experts first demonstrated another solution to the storage problem: do not try to accumulate the energy at all, but feed electricity into a power network as it is generated. This would require a backup power source for the times when wind generation could not meet the need or else a large enough wind-generating system so that there would always be a sufficient amount of electricity produced somewhere in the complex to satisfy customers' requirements.

Among the most imaginative suggestions for capturing the energy of the wind are those of William E. Heronemus of the University of Massachusetts. Heronemus has proposed massive complexes of windmills for such locations as the continental shelf off New England and Lake Michigan. He has envisioned platforms carrying three 2-megawatt generators (each more powerful than the Smith-Putnam machine) or 200 generators of 200 kilowatts each. The platforms could be anchored offshore or be planted in the ocean or lake bottom. Heronemus has estimated that the 200 generator units might cost between $1,268,000 and $1,413,000 each. He also has described towers rising 800 feet above the ground, half a mile apart, with rows of wind machines hanging from connecting cables like those in a great suspension bridge. A unit a mile long, producing 19.2 megawatts, would cost an estimated $2,880,000. Allowing for inflation (and for the federal government to defray development expenses), Heronemus has said that these generators should cost only about half the amount per kilowatt of the Smith-Putnam machine. Systems combining large numbers of these units could generate tens or hundreds of billions of kilowatt hours of power, according to his calculations, without modifying the weather.

Liquid Sun Bearers

Our ancestors early recognized the transport value of running water. They saw that logs or rafts could easily be moved, at least downstream, by the current. The harnessing of those currents, however, took a long time.

Water Mills. The precursor of the water mill was probably the so-called Persian wheel or saqiya. This was (and is) a device for lifting water. A series of buckets attached to the rim of a vertical wheel standing in water scoops the water and, at the top of the turn, tips it into a sluice. It is manipulated by muscle power. Men or animals go around in circles turning a toothed horizontal wheel that meshes with a vertical gear that turns the bucket wheel. That cycle can be reversed: if the weight of water can be lifted this way, the weight of water falling into buckets or onto paddles can drive a mechanism. By 85 B.C. the Greek poet Antipater was praising such a machine for liberating from toil the women who operated the primitive hand mills or querns for grinding grain. The vertical water mill was described by the Roman engineer Vitruvius, about 27 B.C. In China before the first century A.D. a horizontal type of water-driven quern was in use. This consisted of a paddle wheel flat on its side being turned by the current of a running stream or, more effectively, from a well-directed jet from a chute. The vertical spindle went through the hub of a fixed millstone and turned the upper stone to grind the corn.

The vertical waterwheel was disseminated by the Romans throughout western Europe. It appeared in three types: overshot, undershot, and breast. The first stood in the stream under a well-directed chute from which the water hit the bladed wheel. The second stood in the stream with the current flowing underneath and pushing the paddles. The third was a compromise, the stream hitting the wheel at axle level and flowing underneath. When the Goths besieged Rome in A.D. 536 they cut the aqueduct that supplied water to the Roman grain mills. The Roman general Belisarius set up floating mills on the Tiber River. Each consisted of a pair of anchored boats with a vertical wheel between them, driving the millstones on the boats. This improvisation proved so effective that floating mills were employed on the Tigris, Po, Seine, and Danube rivers as well.

The heyday of the waterwheel was the eighteenth century, when hydraulics became dignified as an academic study. Pa-

Running water has been recognized as a valuable source of energy for hundreds of centuries. In a woodcut, dated A.D. *1580, a water wheel manipulated by muscle power provides the power for a mine.*

pers were read at the Royal Society of London, and refinements and substantial improvements were made in the design and operation of waterwheels. The Industrial Revolution was under way. The cottage crafts were being supplanted by machines—the spinning jenny, invented by James Hargreaves in 1765; the water frame, by Richard Arkwright in 1769; and the spinning mule, by Samuel Crompton in 1779. In 1787 Edmund Cartwright invented the machine loom. These had to be housed in factories where power-driven facilities could be available. At first, the waterwheel provided the power. Not for long, however, because James Watt patented the condenser engine in 1769 and by 1782 had converted the reciprocating (back-and-forth) motion into rotary motion so that it could drive the wheels of industry. Such engines soon supplanted waterwheels.

Water Back at Work. Waterpower came into its own again with the development of electricity generation. The hydraulic turbine made it possible to use the power developed by water falling from one level to another with high efficiency and at speeds that permitted use of a reasonably cheap electric generator. It became commercially feasible to develop large power stations at sites far remote from centers of industrial activity and to transmit the energy.

Water may be collected and stored at high elevations and led through tunnels and pipelines to a station at a much lower elevation, the difference in elevation being known as the head. Since the power that can be developed by a given volume of water is directly proportional to the working head, an installation with a large difference in elevation (high-head) requires a smaller volume of water than one with a small difference (low-head) to produce an equal amount of power. The pipelines and turbines required for a high-head are smaller, and the whole installation is cheaper. For these reasons such countries as France, Norway, Switzerland, and Italy—which have mountainous regions, subject to heavy rainfall and in close proximity to industrial regions needing large amounts of electrical energy—are in favorable circumstances for hydroelectric development. Most of the world's waterpower sites that are reasonably close to industrial communities are in regions in which gradients are medium. The aggregate power developed from medium- and low-head plants is far greater than from high-head plants. Low-head plants using heads of twelve to eighteen yards are usually located on rivers in which the gradient is small. Sometimes a natural head is

The huge Cariba Dam in Zambia produces vast amounts of hydroelectric power for the country's copper mining and smelting operations. The Copperbelt mines account for approximately 70% of the electric power consumption in Zambia.

available because of the presence of rapids or waterfalls. In some cases a dam is built.

Where a dam is built, the powerhouse is often built on one flank of the dam with a short headrace or tailrace, and the dam itself is used as the spillway over which excess water is discharged in time of flood. Where the river flows in a narrow steep gorge, the powerhouse may be constructed in the dam itself. Where a river has a long steep bend, it is often possible to cut across the neck of the bend and use the head between the two ends of the cut.

Low-head, propeller-type turbines, inward-flow pressure turbines, and Pelton wheels (a rotor driven by the impulse of a jet of water on curved buckets fixed on the periphery) are used in modern hydroelectric plants. The Pelton wheel runs more slowly than the pressure turbine and is better suited for extremely high heads. The pressure turbine is built in units capable of developing 100,000 kw. Efficiencies of 93% were attained in tests of the vertical shaft turbines at Niagara Falls, and values approximating 90% are common. In a medium-head plant the turbine efficiency may vary from 70% at one-quarter load to 90% at full load.

By the late 1970s about 4% of the energy consumed in the United States came from the generation of electricity by hydroelectric power stations. A few countries, such as Iceland, derive most of their electric energy from this source. Unfortunately, most U.S. hydroelectric installations are built to supplement other sources of electric power and are considered effective only when they are reasonably large. Smaller installations, however, can often be quite effective. By the late 1970s some communities and a few individuals had already made use of these possibilities or were planning small-scale waterpower projects.

In 1977 the U.S. Army Corps of Engineers estimated that about 49,000 dams presently existing on rivers and streams throughout the United States do not have turbines to generate electric power. A sufficient number of these dams could be so equipped to provide an estimated power output equivalent to more than thirty commercial-sized nuclear power stations.

Tidal Power. According to the *Domesday Book* (compiled for William the Conqueror of England in 1087) the Anglo-Saxons of Dover were using the tides to grind their grain. Harnessing the tides as a source of energy has for long been a popular fancy. The costs of doing so, however, along with the environmental impact, have been major deterrents.

Many ideas for harnessing the tides to operate turbines to produce electricity were put forward in the first half of the twentieth century, but no scheme proved technologically or economically feasible until French engineers constructed the Rance power plant in the Gulf of St. Malo in Brittany between 1961 and 1967. A dam equipped with reversible turbines (a series of fixed and moving blades that rotate) permits the tidal flow to work in both directions. The Rance plant has twenty-four power units of 10,000 kilowatts each; about seven-eighths of the power is produced on the more controllable ebb flow. The sluices fill the basin on the incoming tide and are closed at high tide. Emptying does not begin until the ebb tide has left enough head to operate the turbines. The Soviet Union has constructed a tidal plant of about 1,000 kilowatts on the White Sea.

Among the proposed installations waiting for the billions of dollars necessary for their development are the Severn Barrage in Great Britain, the Bay of Fundy (Passamaquoddy) near the U.S.-Canadian border, and the San José Gulf in Argentina. Their common characteristic is the tidal bore,

The Rance hydroelectric plant in the Gulf of St. Malo, France, derives its energy from tidal power. During high tides the dam is opened to allow water to fill the artificial basin behind it; at low tides water is allowed through the dam in the opposite direction. Water flowing in either direction is used to drive reversible turbines.

which advances as a turbulent wall up a river or an estuary. Bores can be more than 3 yards high, with velocities of 5 to 7.5 yards per second.

The environmental reservation about tidal power is the awareness of the critical estuarine phase in marine life, which the installation of power facilities would certainly affect. Whether the taming of the bore would improve or damage the estuarine environment is a matter for debate.

Wave Energy. Considering how for thousands of years man has been tossed about, made seasick, and been shipwrecked by waves, it is strange that until quite recently comparatively little was known about their nature. Before the landings of Allied troops on the Normandy coast of France during World War II, when waves at a given place at a given time on a given day were crucially important, commandos had to make dangerous sorties onto enemy-occupied beaches to make geological samplings, particularly of peat, to see how the waves scoured the deposits and how and where they finished up in the English Channel. Meanwhile, the U.S. oceanographers H. U. Sverdrup and W. H. Munk were developing a series of dimensionally complete empirical relationships on the basis of which it was possible to forecast wave height and period from weather maps. Thereafter, much progress was made to give more complete statistical descriptions of the sea surface.

Whenever wind blows over water, the surface is turned into waves. (More violent waves can be caused by submarine earthquakes, landslides, or volcanic eruptions.) When wind-raised waves travel out of a storm area, they advance as a swell and after traveling large distances become a series of long, low, fairly even undulations, until the swell travels into shallow water. Then a wave crest rises sharply from the water surface until it breaks, and the foaming white surf charges toward the shore and up the beach.

In a wave every molecule of water is moving in a vertical circle. At the surface, the diameter of the circle equals the wave height. The bigger the wave, the bigger the diameter of the system that is revolving. A heavy sea is raised when strong winds blow for many hours over large ocean areas. To produce a wave twenty yards high a sixty-knot wind would have to blow for twenty-four hours over a "fetch" (the distance from the storm center) of about 500 miles.

The transformation of a seemingly orderly swell into steep, violent breakers was at one time thought to be due to friction, but the wartime studies showed that the restricting effect of the seafloor on the circular motion of the wave was the factor. In deep water the circular motion extends to about one-half of the wavelength, the distance between the crests of two succeeding waves. In shallow water the circular motion is squeezed into an elliptical motion (like a crawler tractor). As the waves pile into shallow water there is also a decrease in the wavelength as though the wave train were being compressed like an accordion. The waves themselves change shape. The swell turns into a series of steep crests. The shoreward face becomes concave and the water plunges from the top, striking the water in front with great force.

All the energy that the waves have accumulated from the wind for many hours over long fetches is thrown into the collision with the shore. High breakers can do immense damage. In The Netherlands a breaker lifted a twenty-ton block of concrete 3 yards and deposited it on the pier some 1½ yards above the high-water mark.

The problem is how to convert this liquid solar energy (for that is what it is, from the sun, through the wind, to the sea) into convenient form. The form has to be electricity. And it has to be extracted before it rampages in the form of breakers.

In Great Britain with its wave-girt coastline wave energy has become a major government concern. Five different British research teams with five different plans for generating electricity from waves were active in the late 1970s. An experimental tank, which they could all share, was set up at the University of Edinburgh. It could, with computer control, reproduce every known wave and cross-wave system. It could concoct storms such as had never occurred in nature. Fundamental information could be obtained, and working models could be exposed to the kind of conditions that they would have to survive in the oceans.

As an illustration (and not necessarily as the "preferred system") the Salter Duck may be cited. This is an offshore system proposed by Stephen Salter of the University of Edinburgh. Simplistically, it consists of generators shaped like ducks. It exploits the rotary motion within the ocean swell and employs the waves' energy to generate electricity. The ducks are lined up in an articulated barrage that can be almost any length. It would be moored in deep water at a reasonable distance from the shore and would transmit the electricity by cable to the national grid on the mainland. Any duck can be replaced, thus simplifying maintenance, and the articulated structure (unlike a dam) can respond to storm conditions during which the ducks submerge out of harm's way. Environmentally, it is "clean" energy but it also offers a bonus: the system, having extracted the energy from the waves, provides on the lee side calm waters with many possibilities of sea farming.

The Ocean Hot-Water Cistern. Throughout the eons of history the Earth's ocean has been an efficient collector of solar energy. Radiation from the sun penetrates the water and is trapped, with the result that the ocean surface temperatures in tropical latitudes are maintained at about 77° F (25° C) throughout the year, both day and night. By contrast, the temperature in the ocean depths constantly remains only a few degrees above the freezing point of water. In some areas ocean currents such as the Gulf Stream carry the warm surface waters into the temperate zones, creating a thermal gradient of as much as 36° F (2° C) between water at the ocean's surface and layers only a few hundred yards in depth.

Even though such temperature differences are not large,

heat engines can be built that use this small thermal gradient; this was demonstrated as early as 1929 by the French engineer Georges Claude, who built and operated a small (22 kilowatts) thermal gradient power plant off the coast of Cuba. The efficiency of such a system will never be more than 5%, but the total available energy is tremendous. For example, the total thermal gradient energy passing the coast of Florida each day in the Gulf Stream is about 10,000 times the average daily use of electrical energy in the entire United States. Systems of reasonable size are being considered to tap this source of energy.

The concept was given a new lease on life by the recognition that a closed-cycle power system—using a volatile working fluid, such as propane or a commercial refrigerant—would be appropriate for the temperature level involved. The system would consist of a propane boiler, heated by warm surface water, a propane vapor-driven turbine generator combination, and a condenser cooled by water drawn from a depth of about 1,000 yards.

The potential of a solar sea power plant cannot be fully evaluated until day-to-day operating costs and various technical problems are confronted and solved, but Clarence Zener of the Carnegie-Mellon University in Pittsburgh, Pa., has estimated that such a facility could be built more cheaply than a nuclear power plant or any other type of solar power plant now under consideration. The electricity from a sea power plant, built offshore, might be used to break water down into hydrogen and oxygen. Piped back to the mainland, the hydrogen could be used in a fuel cell to produce electricity. As an example of the power available to a solar sea power plant, a panel estimated that 26 trillion kilowatt-hours per year of energy could be recovered from the Gulf Stream off the southeast coast of the United States, enough to meet the country's projected energy needs in the year 2000.

Harvesting Energy

Nature does a superbly efficient job in converting energy through photosynthesis into living plants and, through them, into living creatures. Nothing that the ingenuity of man has produced thus far can effectively imitate, much less improve upon, that process by which solar energy is stored in what is termed biomass—and that energy could be harvested.

Only a century ago one type of biomass material, wood, was the primary fuel used in the United States, and in many parts of the world wood is still the primary source of heat. Biomass also offers a natural and convenient energy storage system in which energy can be held for relatively long periods of time. Dried plant matter has an energy content by weight that is less than (but usually more than half) that of bituminous coal.

Many uses of biomass as a source of energy have been either proposed or tried, usually on a small scale. One suggestion is tree farms on which fast-growing pines, sycamores, poplars, and other high-yield species would be produced for firewood. Other possibilities include fast-growing annuals, such as surgarcane and sunflowers. Another suggestion, the growing of kelp in the open ocean on huge rafts, would not require use of productive land surfaces. Another interesting photosynthesis system in water is exemplified by a water hyacinth plantation at Bay St. Louis, Miss., which obtains its nutrients from raw sewage, thereby providing sewage disposal as well as energy. These water hyacinths can produce up to ten tons of wet biomass per acre per day, which reduces to about a half ton when dried.

Unlike the burning of fossil fuels, burning of dried plant material should not increase the carbon dioxide content of the air because it is only returning carbon dioxide that was taken from the air a relatively short time earlier during the growth of the plant. A significant factor that limits the growth of plant life and must be considered in land-based biomass systems, however, is lack of adequate water supplies.

In addition to direct burning, energy can be extracted from biomass materials as methanol, ethanol, and various hydrocarbons that are more easily transported than solid material and are often more convenient fuels for some specific uses, such as the operation of moving vehicles. Engineers at the University of Santa Clara, Calif., demonstrated that automobiles can be modified to operate with alcohol as a fuel, and the state of Nebraska successfully tested a mixture of gasoline and ethanol (gasohol) in state-owned vehicles.

The basic idea of the Nebraska project was to look for other markets for the grain grown there. To use agricultural crops as a commercial energy source, however, is not yet economically feasible because the price per calorie received by the farmer for wheat or corn to be used as food is between two and four times the price per calorie of oil. On the other hand, several hundred million tons (dry weight) of organic wastes are produced each year in the agriculture, forest, and

Biomass, which has its origin in plant photosynthesis, offers a natural and convenient system for storing energy from the sun. Certain marine species of brown algae called kelp could be cultivated on large floating "farms" in the ocean and harvested for fuel and other purposes.

other photosynthesis-dependent industries. Many of these wastes are recoverable at a cost per calorie that is less than the equivalent cost of imported oil.

Some types of desert plants appear to be excellent biomass energy converters. During World War II, for example, a desert shrub called guayule was used extensively in the United States as a source of rubber. Another desert plant, jojoba, can produce an oxidation-resistant oil with many industrial and consumer applications. Claims have been made that a substance resembling gasoline can be obtained from at least some species of the genus *Euphorbia.* All these plants can be grown in regions of the southwestern United States where production of other crops often is not feasible.

In the well-known bacterial process of anaerobic digestion, which is presently in use on a small scale on a number of farms throughout the world, animal manure is turned into methane. Methane from the breakdown of urban sewage has also provided the energy requirements for many sewage disposal plants. Yet it is not known exactly what fraction of existing animal wastes can be collected and used. Effective use of this resource appears to depend on the scale. If a large number of small units are put into operation on individual farms the amount of energy that can be produced will be much larger than that from large units built to convert large quantities of animal wastes (such as from feedlots) to methane.

Many other ways also exist for the conversion of biomass to useful energy sources. At present the extent to which such sources can replace conventional fuels is not known, but the possibilities appear to be large. Another uncertainty is cost, but, as with all other new developments, initial costs are usually not indicative of final costs; after a product is more widely used, mass production of equipment and supplies becomes possible.

Direct from the Sun

The world, we are told, is short of energy. How so? Every twenty-two minutes as much energy strikes the Earth in the form of solar radiation as is used in a whole year by all consumers throughout all the world. In two weeks this solar energy exceeds all known coal deposits.

The total amount of solar energy reaching the Earth is about 7,000,000,000,000,000,000 (7×10^{18}) kilowatt-hours per year, more than 30,000 times as much as is used in all man-made devices. Much of this, however, has to be discounted in terms of availability. First, the source is cut off at night. Second, there are curtains of clouds and haze. Of the solar energy intercepted by the Earth about 30% is reflected from clouds and polar ice back into space. About 47% is absorbed at or near the Earth's surface and is transformed into heat, which is vital to the support of life, and about 23% goes into the evaporation of water, both from seawater and through transpiration from the leaves of green plants. This water ultimately returns to the Earth as rain or snow. Although these three categories account virtually for 100% of the sun's radiant energy, two other categories are very important, even though fractionally small. About 0.2% is used to produce winds, ocean waves, and other phenomena resulting from atmospheric motions. An even smaller amount, 0.02%, is absorbed by plants in the process of photosynthesis—the chemical transformation that must take place for living things to exist on the Earth.

Man has sought for many centuries to make direct use of the sun's radiant energy. Thus far, his success has been meager, primarily because sunshine is relatively low in intensity. Apparatus intended for use with solar energy must in general be large in area, movable, so that it can follow the sun's apparent motion, and provided with some means for storing energy for use when the sun is not shining. Despite these formidable difficulties significant advances have been made in using solar energy to generate heat, to bring about chemical reactions, and, most recently, to produce usable amounts of electricity.

Man learned long ago that heat is produced when the sun's rays are absorbed by a blackened surface. He found, too, that the amount of heat thus produced could be greatly intensified by using reflection or refraction to concentrate a large area of sunshine onto a small target. The concave silver mirrors left behind by the Incas and the convex quartz lens found in the ruins of Nineveh were probably used to light sacred fires by means of concentrated sunbeams. The ancient Greek mathematician Archimedes is said to have devised a battery of mirrors to defend Syracuse in 212 B.C. by burning the sails of an invading fleet "at the distance of a bowshot." In 1747 the French naturalist Comte Georges Buffon set up a group of 140 flat mirrors in a Paris garden and ignited a stack of wood placed about 200 feet from his reflectors, thus demonstrating that the feat attributed to Archimedes might indeed have been accomplished.

French chemist Antoine Lavoisier, who began to use solar energy for scientific purposes as early as 1772, was probably the originator of the device that is known today as the solar furnace. Enclosing specimens of various substances in transparent quartz vessels, he placed them at the focal point of a lens fifty-two inches in diameter and used concentrated solar radiation to heat them in a partial vacuum and in controlled atmospheres of oxygen and other gases. He first called attention to one of the principal virtues of the solar furnace when he wrote, "the fire of ordinary furnaces seems less pure than that of the sun."

Because of its unique ability to heat materials by intense radiation alone for long periods of time without contamination the solar furnace has come into prominence as a tool for high-temperature research. Scientists have found that parabolic reflectors taken from antiaircraft searchlights can concentrate solar radiation so effectively that temperatures as high as 6,100° F (3,400° C) can be attained. A parabolic reflector is a concave mirror that has the form of a paraboloid of revolution. This causes all rays emanating from the geometrical focus and reflected from the surface to be parallel to one another and to the axis of symmetry and rays from a distant source to be reflected to the focus. Most of the large solar furnaces avoid the inconvenience caused by target movement by using a flat mirror called a heliostat to track the sun across the sky and reflect its rays into a fixed paraboloidal concentrator. Among the largest furnaces of this type are those at Mont-Louis in the French Pyrenees and at Natick, Mass. Both use concentrators that are more than thirty feet in diameter, made of many small curved glass mirrors arranged approximately in a paraboloidal configuration.

Solar steam generators were built by many pioneers during the nineteenth century. All of these used arrangements of movable mirrors to concentrate large amounts of solar radiation upon blackened pipes through which water was circulated and turned to steam. Both steam and hot-air engines were operated in this way with some degree of success as early as 1870, and ice was produced in Paris in 1878 in an ammonia-absorption refrigerator operated by a solar boiler. The use of solar energy for pumping irrigation water has been tried in Arizona and California, employing a number of solar pump-

A solar generator utilizes energy captured from the sun to produce electricity at Odeillo-Via in the French Pyrenees.

ing stations using conical concentrators thirty feet in diameter.

The largest of all solar power installations was erected in 1912 on the bank of the Nile River near Cairo, Egypt, using a total of 14,000 square feet of concentrating surface in the form of seven parabolic troughs, each 205 feet in length. The 100-horsepower steam engine connected to this great solar boiler actually produced between 50 and 60 horsepower continuously during one five-hour test, but the system was not economically competitive with other pumping apparatus and it was abandoned during World War I.

Trapping the Sun's Heat. All of these early installations suffered from the same deficiencies—irregularity of operation and excessive cost—and none survived. Interest in solar power revived with the advent of artificial Earth satellites, since these applications avoid such earthly problems as clouds and atmospheric absorption. Serious consideration is again being given to furnaces that would receive the full complement of solar radiation, 1.4 kilowatts per square meter, whenever their collectors turn toward the sun, and reject unused energy by radiation into space.

When heat is needed at moderate temperatures for such

purposes as the distillation of saltwater, the growing and drying of agricultural products, cooking, and the heating of buildings and domestic hot-water supplies, solar energy can be used with considerable success, thanks to the "greenhouse" effect. Common window glass and many plastic films have the property of transmitting nearly all of the solar radiation that falls on them, but they are relatively opaque to the long waves of radiation emitted by any surface heated to 212°–392° F (100°–200° C). A simple blackened wooden box covered with several sheets of glass or plastic film therefore acts as an effective trap for solar energy when it is turned toward the sun. If its sides and bottom are insulated, the internal temperature can be raised as high as 302° F. (150° C) by the unaided rays of a bright sun. M. K. Ghosh in India (1945) and Maria Telkes in the United States (1955) improved this simple solar stove by adding reflective wings to direct more sunshine into well-insulated ovens.

Solar water heaters have been in relatively wide use in Florida and the southwestern United States for many years. Most of these consist of flat sheets of blackened metal provided with tubes through which water circulates by natural or forced convection whenever the sun is shining. Wastage of the absorbed solar energy is minimized by covering the upper surface of the metal with a glazing of glass or plastic film, separated from the metal sheet by a small air space. The lower surface is insulated with rock wool, vegetable fibers, or a reflective coating that cuts down radiation. Heaters of this type do not have to follow the sun's motion because they operate effectively if they are mounted in a fixed position facing the south (in the Northern Hemisphere) and tilted so that their angle with the horizontal is equal to the local latitude plus 20°. Temperatures up to 140° F (60° C) can readily be reached on sunny winter days, and the domestic hot-water requirements of a typical family living in regions south of latitude 35° N can be met during most of the year by a heater having as little as forty square feet of surface.

Similar but much larger collectors can supply most of the heat needed in homes of moderate size in favorable localities. Storage of heat for use at night or during cloudy periods is accomplished by using large insulated tanks to store water heated during daylight hours. This system is particularly effective when radiant heating is used, with the warmed water flowing through tubes in floors and ceilings. Air can also be

A solar energy system installed in an individual home can supplement or supplant conventional space heating and hot-water systems. In this particular case, solar reflectors constructed as troughs with parabolic cross sections focus sunlight onto fluid-filled tubes suspended along the focal line of each reflector.

heated in suitably designed collectors and used in conventional warm-air systems. Heat storage can then be provided by using beds of gravel or containers filled with chemicals such as Glauber's salt that absorb heat at a suitable temperature when they melt and give off the stored heat when they solidify.

An important step in reducing the use of fossil fuels should occur if residential space heating is transferred to solar energy, because home heating represents about 13% of all energy use in the United States. Heating of commercial buildings by solar energy is also technically feasible, and appropriate systems have already been installed on a number of such buildings. The New Mexico Department of Agriculture Building in Las Cruces, for example, has derived all hot water and space heating from solar energy since it was completed in 1975. Several buildings in more northerly climates have also been fitted with solar energy equipment that provides a large fraction of their winter heating requirements.

Air conditioning in summer is not quite as easy as space heating in winter, but techniques already exist to accomplish this task, based on principles used in early refrigerators that employed natural gas as their energy source. For example, the New Mexico Department of Agriculture Building receives between one-third and one-half of its summer cooling by using a system that produces chilled water, which, in turn, is used to cool the air circulated throughout the building.

Solar radiation can be used for many purposes other than simple heating of air or water. Solar energy water pumps, for instance, have already been built and operated. The pioneer work in this field was concentrated in West Africa, but in the late 1970s it was also becoming important elsewhere.

Solar Cells. A solar cell is a device that converts light rays into usable electricity, as was demonstrated throughout the U.S. and Soviet space programs. It does this by generating an electromotive force as the light falls on regions of transition between two semiconducting materials.

The reasons usually offered for minimal use of these cells in ground-based applications are cost and efficiency, since cost per kilowatt using solar cells is about twenty times the cost of power from central generating stations. Technological developments, however, continue to lower the cost of solar cells and, if the trend continues, probably will make them economically competitive near the end of the twentieth century. Even though cells now commercially available convert only about 10% of the incident sunlight to electrical energy, experimental cells have been constructed with much higher efficiencies—silicon cells at 19% and gallium arsenide cells at 24%.

Because solar-cell output is directly related to incident light intensity, the use of light concentrators reduces the cost per kilowatt for a single cell. Furthermore, heat generated by the 90% of the unconverted solar energy can be carried away by a circulating fluid to perform other tasks.

An experimental array of 135 silicon cells and concentrating lenses at the Sandia Laboratories near Albuquerque, N. Mex., produces one kilowatt of electricity directly from sunlight. Excess heat piped from cells by liquid coolant can be put to useful work. The U.S. Department of Energy's Solar Thermal Test Facilities at Sandia consist of about 300 heliostat arrays, each of which contains twenty-five four-foot-square mirrors. As many as 6.5 million watts of solar thermal energy can be tested there.

Even with significant cost reductions, power plants that draw electricity from solar cells could not be operated at

Heliostats, or sun-tracking mirror arrays, are being tested at the Solar Thermal Test Facility at Sandia Laboratories in New Mexico. The heliostats are designed to focus concentrated sunlight onto a test boiler mounted on the top of a 200-foot "power tower."

night or on cloudy days. Electrical power, therefore, must be stored or drawn from some other source during the "down" period. As an alternative, Peter Glaser of Arthur D. Little, Inc., in Cambridge, Mass., visualizes huge satellite solar power stations that would orbit the Earth at an altitude of 22,300 miles and be exposed to the sun nearly twenty-four hours a day.

In Glaser's plan electrical power in such a station would be produced by a battery of solar cells ten square miles in area and surrounded by reflecting mirrors. The electric current would then be converted to a beam of microwaves, transmitted back to a receiving antenna on the Earth, and reconverted to electricity. Glaser has estimated that satellite solar power stations, which would weigh approximately 5 million pounds, would generate 10,000 megawatts of electrical power each year, equivalent to the annual energy requirements of 10 million people.

Glaser's high-flying idea generates new problems, however, one of which is that to lift the many components of the power station into orbit about 500 two-stage trips of the space shuttle, which is scheduled to go into operation during the 1980s, would be required. This, of course, would cost billions of dollars.

12.
Geothermal Energy—
Power From the Nether World

The formation of geothermal regions within the Earth and their location and size in the United States; methods of harnessing geothermal energy and the feasibility of its commercial production, including cost and environmental factors; and the gasification of coal and MHD

Old Faithful has not spouted in vain. Since time beyond memory the geyser has been erupting every hour, jetting water for five minutes at a time to as high as 150 feet. This not only edifies tourists in Yellowstone National Park, Wyo., but it also, like its less regular neighbors and the giant geysers of Iceland and New Zealand and boiling mud pools, noneruptive hot springs, and curative spa waters all over the world, reminds us that there is energy below the Earth's surface.

The natural thermal energy of the Earth's interior is called "geothermal energy," and it requires a great deal more attention than it has received thus far. It ties up neatly with the exciting new discipline of plate tectonics, which has to do with the drifting continents being carried on rafts that are the broken shell of the Earth's crust, the so-called plates. Those plates sometimes jostle each other, sometimes separate and expand the ocean floor, and sometimes dip one under another, creating the deep-sea trenches. If a plate moves away, hot rock oozes out from the interior to calk the gap. That is what happens at the mid-ocean ridges and the continental rift valleys. When one plate dips under another, the sinking plate gives rise to deep-seated earthquakes and the friction it creates causes volcanoes to erupt on the far side of the trench. By such evidence scientists have determined the configurations of the plates and where the action is likely to occur.

Intense heat has been built up within the Earth over the billions of years of accumulation of the thermal energy remaining from the decay of natural radioactive elements and from the frictional dissipation of energy released by the gradual segregation by density of materials within the Earth's interior. This white-hot interior is protected by the excellent insulating qualities of the rocks of the Earth's mantle and crust. The thermal conductivity of most rocks is so low that they have kept the molten core and the mantle of the Earth from cooling, even over the eons of geological time. The long history of production of the Earth's internal energy and its escape to space has resulted in the establishment of natural heat flow throughout the globe. The heat flow unit (HFU) is one microcalorie per square centimeter per second. The average heat flow over the continents and the oceans ranges from 1.7 to 1.9 HFUs. This average is remarkably constant over nearly half to two-thirds of the Earth's surface.

There are, however, areas where natural heat flow is substantially higher, often by more than a factor of ten, than the global average. Examples include areas of volcanoes and rift valleys where molten rock rises from deep within the Earth's interior to the surface. In these regions molten rock pushes its way upward, rising buoyantly and transferring large quantities of mass and thermal energy from deep within the Earth's mantle toward the surface.

The heat is often transferred to water in the subsurface, and the energy content of this water can be recovered. If the natural pressures are low the water may boil underground and steam may permeate the natural porosity of the rocks. A well drilled into one of these steam zones produces large quantities of geothermal steam. This steam can be fed into turbines and used directly to generate electricity. If the pressures in the water are sufficiently high, boiling is prevented and large quantities of hot water accumulate in the subsurface. This hot water, on flowing into a well, will quickly boil, producing much the same effect as a bottle of soda water being uncorked, and the mixture of steam and boiling water will flow to the surface with great force. The steam can be separated and utilized in turbines in the same manner as with dry steam. Under certain circumstances the hot water can be used directly without boiling, and this promises to become a new and major source of useful energy.

Recognized for thousands of years, geothermal energy has been used since man first discovered boiling springs and used them for hot baths. More recently, at the start of the twentieth century, natural steam began to be harnessed for the generation of electricity; by the 1970s many countries throughout the world were producing electricity from this source. The rapid rise in the cost of fossil fuels and the recog-

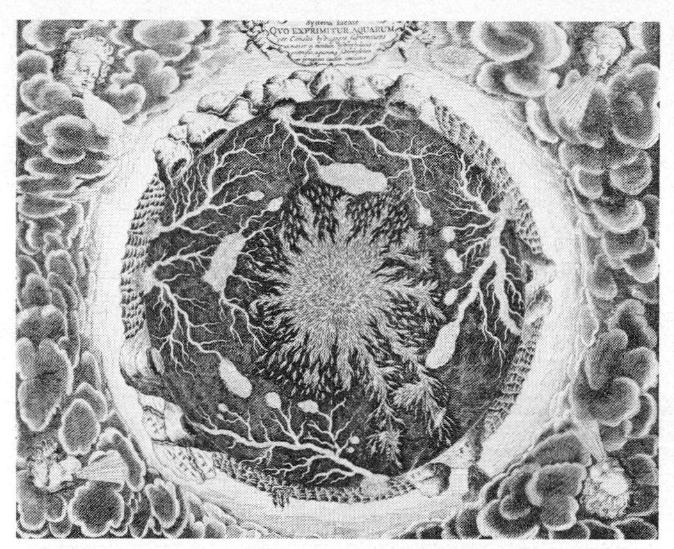

Seventeenth-century concept of the Earth's interior (top) is depicted in the engraving "Concerning Volcanic Activity," which appeared in the book Mundus Subterraneus. *Steam, which can be fed into turbines to generate electricity, rises from various areas of the Earth's surface (bottom).*

nition that they pollute the atmosphere has caused an increased awareness among scientists of the potential of geothermal fields. Though geothermal energy is primarily considered in terms of the heat content of subsurface water, an even larger amount exists in hot rock where little water is present. Technology is being improved to develop commercially the energy in hot, dry rock.

Formation of Geothermal Regions

One process for rapidly transferring heat to the surface occurs in areas of rifting. One of the important areas of geothermal activity is Iceland, which lies on the crest of the mid-Atlantic ridge. Along the Great Rift in East Africa there is geothermal potential in such countries as Kenya and Ethiopia. Another area of rifting is the Gulf of California and the Mexicali and Imperial valleys at the head of the Gulf; there the East Pacific Rise has caused a rift valley to form, splitting that part of California west of the rift and its associated San Andreas fault away from the rest of North America.

In rifts there appear to be deep zones of tension that penetrate hundreds of miles into the mantle, fracturing it and releasing pressure. Molten basalt from partially melted rocks

In Iceland, one of the world's major sites of geothermal activity, steam pours from fissures in the Earth's surface. The island lies along the crest of the mid-Atlantic ridge, where rifting causes heat to be transferred rapidly to the Earth's surface.

of the Earth's mantle flows into these fractures and then squirts to the surface with flow rates approaching nearly the speed of sound, according to some investigators. In some cases this basaltic lava will flow out at the surface, while in others it may be frozen by contact with subsurface water and never appear through the overlying cover of wet sediments. The thermal energy content, or enthalpy, of the molten basalt, which rises from the mantle at temperatures between 1,800° and 2,000° F (982° and 1093° C), is, however, transferred to the overlying waters. In this way massive amounts of thermal energy are rapidly pumped from several hundred miles within the mantle to the surface of the Earth.

Spreading of the Earth's crust can also take place without any actual large-scale break in the surface. This results in a process of crustal thinning, or necking, and is accompanied by an upward rise of the rocks of the mantle. The hot rocks of the mantle, therefore, come much closer to the Earth's surface in an area of crustal thinning, and there is an increase in heat flow in such a region. This increased flow can be trapped by the insulating qualities of shale in sedimentary basins. Consequently, the water in large sedimentary deposits formed over an area of crustal thinning can have extremely high temperatures. This hot, high-pressure water is called geopressured geothermal water. It is abundant in sedimentary basins such as the Gulf Coast, where it occurs in large quantities offshore and onshore in Texas, Louisiana, Mississippi, Alabama, and northern Florida. It is also found in California and in many places in Africa and the North Sea.

The potential for a geothermal energy source is also high in an area where two crustal plates impinge. The downward-plunging crust carries large quantities of sediment and water with it. This rock-water mixture has a substantially lower melting temperature than is normal for mantle rocks, and, consequently, the process transfers large quantities of rock material into the mantle in an environment where it should be liquid instead of solid. The molten material appears to accumulate in gigantic "droplets" measuring miles to tens of miles in diameter, which then rise buoyantly, melting their way through the overlying mantle and crustal rocks. These gigantic droplets, when crystallized, constitute coarsely crystalline bodies known as batholiths and stocks. If one of the bubbles breaks through to the surface, it will feed a volcano.

The thermal energy carried from within the mantle to the surface by large masses of molten rock provides an enormous reserve of geothermal power. By the early 1970s scientists understood how to tap that portion of the energy that is present as steam and could utilize in part that portion of the energy present in hot water. The next step will be to get energy from the molten rock itself.

Geothermal Supplies in the United States

The resource base for the geothermal energy potential in the United States consists of all of the heat in the ground that theoretically can be recovered. This base is estimated to be exceedingly large, far larger than the conceivable energy requirements of the country for many thousands or even tens of thousands of years. The size of the total resource base, however, has relatively little meaning in itself. What is meaningful is the amount of energy that can be recovered through technology that is either presently available or may reasonably be expected to be available in the near future, and also the cost of recovering this energy.

The reserves that are known are relatively small, probably equivalent to approximately 38,000 megawatt-centuries of electricity. (A megawatt-century is a megawatt of electricity produced continuously for a century.) By comparison, the probable geothermal reserves are at least 6 million megawatt-centuries, while the undiscovered reserves are believed to be larger than 40 million megawatt-centuries. (Undiscovered reserves are those reserves estimated for known prospective areas by means of geophysical techniques and subject to a discount for the risk factors involved.) This broad range of geothermal potential has to be considered both from the perspective of the degree of uncertainty as to the existence of the reserves and as a function of the cost of energy production in mills per kilowatt-hour. As the market price for energy increases, a larger and larger proportion of the resource base becomes economically feasible to be used.

Because of the interaction of exploration risk and market price there is no simple answer to the question of the development potential of U.S. geothermal reserves. The probable reserves, if they were developed quickly and at a cost that would be competitive with that of nuclear- and fossil-fuel energy sources, would be sufficient to meet all of the electrical generating needs of the country within the next thirty years, provided that they were found within 500 to 750 miles of their market area. Unfortunately, this is not presently the case for the northeastern and north-central portions of the United

States, and for this reason present geothermal energy technology does not appear to be capable of meeting all of the electrical energy requirements by the year 2000. If the technology, however, to exploit the hot, dry rock geothermal potential were to be successfully developed, the price of electricity at the generating site in the northeastern United States would range from twelve to sixteen mills per kilowatt-hour, compared with the cost of electricity from oil and nuclear sources ranging from twelve to twenty mills per kilowatt-hour. Coal prices for environmentally acceptable grades are rising rapidly, and electricity from this source appears to range from twelve to eighteen mills.

There seem to be sufficient undiscovered geothermal reserves of hot, dry rock to meet the total need for electricity for the entire United States for many thousands of years. But until the hot, dry rock technology is achieved it is impossible to base the country's energy strategy on the potential of geothermal energy. It is evident, however, that for the Gulf Coast and the western United States geothermal energy is a leading contender for electricity generation, probably for at least the next thirty-year period.

Harnessing Geothermal Energy

The earliest development of geothermal energy involved creating ponds from natural hot springs and piping the waters to baths. This was followed by drilling shallow wells in areas of natural hot springs in order to increase or stimulate the natural water flow. These wells were often successful, and many former hot springs were replaced by hot flowing wells. Deeper drilling resulted in the production of fluids at higher and higher temperatures.

As more research was done it became evident that there were different kinds of geothermal resources and that each one involved a different technology for energy extraction. Dry steam geothermal energy at the Geysers Field in Sonoma and Lake counties in northern California, for example, is produced by drilling with an oil-well type of rig into fractured metamorphic, sedimentary, and igneous rocks of the Franciscan formation of Jurassic–Cretaceous age.

At the Geysers, steam is produced from natural fractures in the rock. The wells have two valves set on the casing head, and a pipe connects the wellhead with a separator that removes bits of rock, grit, and debris from the flowing steam. The steam then passes the main generating station, where

Electricity is commercially produced from geothermal energy at the Geysers Field of northern California. Steam drawn from natural fractures in the rocks passes through the main generating station after bits of rock and debris have been removed from the flowing steam by a separator.

electricity is produced. The steam turbines are rated at 55 megawatts each, and two turbines are usually housed together in one module in the generating plant. The newest approach is to combine two turbines with a single generator, with projections calling for two 70-megawatt turbines driving a single 140-megawatt generator.

Commercial Production

The United States in the late 1970s was the second largest producer of electricity from geothermal energy and was the country that was expanding production most rapidly. All of the commercial production was at the Geysers Field in northern California, where 520 megawatts of power were in production, sufficient to meet the needs of a city of half a million persons. The second largest area under development in the United States was in the Imperial Valley of California, where several geothermal fields had been discovered and were undergoing test development. Additional fields were discovered in the Long and Surprise valley areas of California, and exploration was occurring throughout the states of California, Oregon, Washington, Idaho, Utah, Nevada, Arizona, New

Mexico, Colorado, Montana, and Wyoming. A major geothermal field is in Yellowstone National Park, where both dry steam and hot-water geothermal resources are abundant. The national park status of the Yellowstone area, however, protected this field from commercial development.

The world pioneer in geothermal development was Italy. Substantial large-scale geothermal development began early in the twentieth century in the area of Larderello; this was later expanded into other areas so that by the early 1970s geothermal capacity in Italy totaled 391 megawatts. This total rose later in the decade. Italian production focused on the development of dry steam, and a major exploration program was under way throughout the country to increase the number of commercially developed geothermal producing areas.

For many years the second most active country in the world in the development of geothermal energy, New Zealand recently slipped to third behind Italy and the United States. The New Zealanders discovered at Wairakei that they had a major hot-water geothermal field for which no producing technology existed. They then proceeded in the 1960s to pioneer in the development of such technology there and at such other fields as Kawerau, Rotorua, and Broadlands.

The widespread use of hot springs for baths caused the Japanese to become among the world's most vigorous utilizers of geothermal energy, even if on a small scale. The presence of about 5,000 hot-spring spas developed for recreation in Japan, however, actually served to significantly inhibit development of many potential hot-spring areas for their power potential. Installed capacity of dry steam at the Matsukawa field totaled 20 megawatts in the late 1970s, and another 30 megawatts were developed in the hot-water geothermal area at Otake.

Iceland has used geothermal energy on a large scale, but only a small fraction has been utilized for electricity production. Instead, the geothermal hot water has been used directly for heating. The capital city of Reykjavik solved its smoke and smog problem when it shifted from burning peat to using geothermal hot water.

Abundant geothermal resources occur in many places in the world where there is insufficient thermal energy content in the water to produce a large quantity of steam by boiling or where the waters may contain chemical constituents that are precipitated on vaporizing. For this reason, a means has to be used to transfer thermal energy through a heat exchang-

At Wairakei field in New Zealand, steam that has been separated from hot water is transported through pipes to a power plant where it will be used to generate electricity.

er to another working fluid, such as freon, or to a hydrocarbon, such as isobutane. (A heat exchanger is a device designed to transfer heat from one fluid to another without allowing them to mix.) This secondary working fluid would then run through a special turbine, which, in turn, would drive an electrical generator. By appropriate selection of materials and components the binary fluid system can operate with much smaller turbines than can a steam system, thereby allowing the turbine system economies to compensate for the cost of the heat exchangers.

A successful binary fluid system was developed at Paratunka, Kamchatka, in the Soviet Union, utilizing water at a temperature of 175° F (79° C) and freon as the working fluid. The principal problem proved to be leakage of freon from the system. Because of the high cost of freon this was an important economic constraint, but once this problem was solved the plant operated successfully. The value of this pilot plant demonstration was substantial because hot waters in the 175° F temperature range are distributed throughout the world.

Geothermal water of the geopressured type like that in oil wells is found in many sedimentary basins. On the Gulf Coast of the United States there are large quantities of salty water occurring at depths of about 6,000 to 30,000 feet (1,829 to 9,144 meters). The temperatures of these hot waters range from 300° to 500° F (148° to 260° C), and they contain substantial quantities of methane in solution. The technology exists for accomplishing separately the recovery of mechanical energy from water at high pressures, the recovery of thermal energy from water at the above temperature, and the separation of methane from water.

Cost Comparisons and Environmental Impact

Electricity produced from steam at the Geysers Field proved in the late 1970s to be the least expensive new source of energy within the Pacific Gas & Electric Company utility service area. Geothermal energy at 90% load factor (ratio of average to maximum load) produced electricity at a cost of 6.43 mills per kilowatt-hour. Oil and gas as an energy source yielded electricity at a price of 11.55 mills per kilowatt-hour. The cost of electricity derived from Wyoming coal used in California was 10.52 mills per kilowatt-hour, delivered within the service area. The projected costs for a nuclear power plant at Mendocino, Calif., were 9.26 mills per kilowatt-hour.

It is evident from these data that geothermal energy represents the least expensive source of new power available within areas where dry-steam geothermal resources are known to be present. The problem is to expand the number of known dry-steam geothermal fields by making exploration and production more attractive. The major stumbling block is that the majority of new dry-steam prospects are on federal lands that are not available for leasing, although a federal geothermal act of 1970 was designed to eventually open some federal lands for such development.

The environmental impact of geothermal energy is almost entirely at the site of the field. The thermal energy dissipated in the course of production of electricity is released to the atmosphere by the evaporation of cooling water. This water is derived from condensation of the geothermal steam itself. Consequently, a geothermal power plant requires no external source of water, and any excess water produced in the operation may be injected back into the steam field in order to minimize ground subsidence and to conserve water. Excess chemical components, such as boron and hydrogen sulfide, are dissolved in the condensate water and reinjected into the geothermal zone, thereby eliminating any surface disposal of chemical-containing waters. Thus, there is no environmental hazard from brines produced from geothermal systems. A small amount of hydrogen sulfide, however, does escape from the evaporative cooling towers in a geothermal plant. In the United States efforts are being made by electrical utilities to devise techniques for inhibiting the escape of this noxious gas, and researchers expect that within a few years the situation will be under control.

An early problem in geothermal development was the noise produced during the testing of steam wells. The extensive use of mufflers, however, has reduced the noise problems substantially. During actual commercial production the steam wells generate such low levels of noise that they are not audible beyond a distance of about 600–3,000 feet.

Little information is available concerning the seismicity associated with geothermal steam production. Theoretical and experimental work on the mechanisms of the generation of earthquakes indicates that the presence of hot water along a rock fracture system tends to cause rock to move by creep and not by the process known as stickslip, which results in seismic shocks. This suggests that an area capable of producing geothermal steam is not likely to be one that generates earthquakes, and also that the process of bringing heat from

Salt brine brought up with natural steam from beneath the Earth's surface is condensed in a tank at Geysers in California.

the Earth's interior up to the surface actually tends to diminish shallow seismic hazards in that particular locale.

The main environmental impact of a geothermal plant is simply its physical existence, which involves not only a cluster of wells but also associated service roads and facilities. This type of impact is remarkably modest considering the massive effect of almost all alternative energy sources. Research and development in the United States, Mexico, and other countries is demonstrating that more and more of a geothermal plant can be placed either entirely below ground or at least partially buried in order to minimize the visual impact.

The relatively low cost of producing geothermal energy, compared with the cost of producing energy from competitive sources, and the remarkably small environmental impact combine to make geothermal energy one of the most attractive new energy sources available. If the technology for exploiting the widespread fields of hot, dry rock can be developed, geothermal power can make a major contribution to the supply of power throughout the world.

Flaming Electricity

Another way of getting useful energy from the nether world would be a combination of subterranean gasification of coal

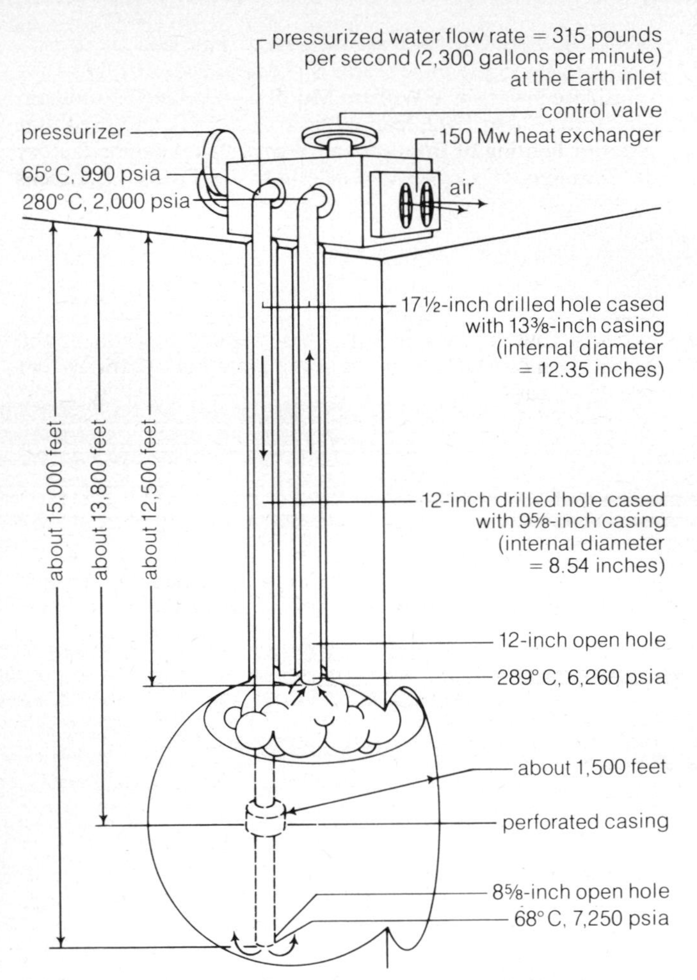

In a proposed method of generating power from the hot, dry rock beneath the Earth's surface, wells are drilled into the rock, which is fractured hydraulically so that two wells are connected. Cold water is then allowed to flow down one well into the rock, where it is heated, causing it to rise up the second well. In a heat exchanger the heat from the water is transferred to isobutane, which then vaporizes and is used to drive a turbine and thereby generate electric power. (Mw=megawatt and psia=pounds per square inch absolute.)

with magnetohydrodynamics (MHD). The idea of turning coal into gas in the mine seams is not new. Indeed, it has been considered ever since William Murdock produced illuminant gas from coal in 1792, learned how to store it, and used it as exterior lighting of Boulton and Watt's steam engine factory at Birmingham, England, to celebrate the Treaty of Amiens in 1802. If coal could be gasified, why hew it out of the ground, haul it to the surface, and carbonize it there? The answer obviously is not as simple as the question because, with many attempts, a really effective way of firing the seams, managing the combustion, and extracting the gases has not yet been found. The solution to this problem might lie in the application of MHD, a method of generating electricity that involves heating a gas at 4,000° to 5,000° F (2204° to 2760° C), at which point the gas becomes ionized (electrically charged). When driven past a magnetic field, the gas becomes a conductor system. The fuel is burned (along with a "seeding" of potassium crystals to raise the level of conductivity) in a long combustion chamber. Under pressure a supersonic stream of gases rushes down a tube that is ringed by a superconducting magnet. The electric current carried by the gas is drawn out by a series of electrodes. Part of the hot exhaust gas is fed back to the burner and part to an air turbine to produce more power. An electrostatic precipitator recovers the potassium "seed" for reuse and also removes all particulate matter. A chemical unit recovers usable nitric and sulfuric acids, and the ultimate exhaust consists of clean gases, carbon dioxide, and nitrogen. Scientists hope to increase efficiency in fuel-to-power conversion from 40 to 50% or possibly even 60%. Not only would this conserve nonrenewable resources but it would also minimize pollution.

MHD has been extensively studied in nearly all industrialized countries. The Soviet Union demonstrated a large-scale unit using natural gas for fuel. Coal, however, would be the preferred fuel because it is the only fossil fuel with reserves large enough to last during the time MHD would be most useful. If gasified underground, so much the better.

When MHD becomes commercially feasible, thermal pollution would be reduced or eliminated. Air pollution would be reduced per unit of electricity generated. There are other bonuses: less fuel would be needed for each unit of electricity generated, and so less land would be disturbed by mining; less water would be polluted by acid drainage from mines; and there would be minimal solid wastes in the form of slag or ash.

Afterword: The Politics of Energy

With the flick of a switch, the modern industrial worker can summon electrons to do prodigious amounts of work no human being would even attempt. In his machine shed, today's farmer has many more "slaves" than were ever owned by the biggest plantation owner. The ordinary housewife has, in her domestic gadgets, more servants than the most ostentatious aristocrat employed fifty years ago. The average car has about two hundred horses safely stabled under its hood.

This energy-at-work (power) provides the index of the wealth of any country and helps to explain why the per capita income of the United States is a hundred times higher than that of India, where power still mainly depends on the muscle power of man or beast. This is the most expensive power, since energy in the form of food calories of even the poorest diet costs twenty times more per unit than the electricity generated by nuclear reactors.

The often-heard phrase the "energy crisis" is the alarmed awareness of the dependency of a machine civilization on preempted forms of energy—preempted because one form of energy has been preferred to the exclusion of others. A boat without oars (muscle power) and without sails (wind power) is helpless when the engine runs out of gasoline (fossil fuel).

There have been plenty of examples of this dependence. During World War II Adolf Hitler's hopes of success depended not on the muscle-powered infantry divisions but on his air force and tanks, which required oil. Germany, however, had to import its oil supplies, with an important exception: before beginning the war the Germans developed the Bergius process for converting coal into oil. In a succession of stages high-grade gasoline could be produced.

The main supply of fluid fuel for the German war machine was from the oil fields of Ploesti in Romania, and a desperate object of Hitler's eastern strategy was to secure the Baku oil fields in the U.S.S.R. His thrust into the Caucasus had failed in 1942. In May 1944 Ploesti was bombed by the U.S. Air Force. After that the internal combustion engines of Germany depended on synthetic fuel, but the outcome was a foregone conclusion. Of the sequel, Albert Speer, the German armaments minister, wrote:

> *I shall never forget the date May 12 [1944]. On that day, the technological war was decided. Until then we had managed to produce approximately as many weapons as the armed forces needed, in spite of their considerable losses. But with the attack of 935 bombers of the American Eighth Air Force upon several fuel plants in central and eastern Germany, a new era in the air war began. It meant the end of German armaments production. The next day, along with technicians of the bombed Leuna Works [the main coal-hydrogenation plant] we groped our way through a tangle of broken and twisted pipe systems. The chemical plants had proved to be extremely sensitive to bombing . . .*

In December 1944 the Germans launched the Battle of the Bulge. The Ardennes offensive, which took the Allies by surprise, was fueled by synthetic oil. The objective was Antwerp, Belgium, not only to split the Allied forces but also to overrun their fuel depots and supplies for the Rhine crossing and to replenish the German tanks and air force with captured oil. It failed. With Leuna and the synthetic-oil plants damaged in the rear and the offensive blunted at Bastogne, the massive German armor ground to a halt like bog-bound dinosaurs.

On Nov. 9, 1965, power failed in the northeastern states of the United States, and an area containing about 30 million people was blacked out. Subways stopped, and 800,000 people were stranded at the rush hour. Traffic lights failed. Neon signs went out. Hot dogs cooled in microwave ovens, and chickens idled on spits. Motion picture and television screens were blank. Printing presses stopped so that they could not print the big news story of which they were a part. Elevators stopped between floors of skyscrapers. Sewage pumps, water pumps, air conditioners, and refrigerators failed. Airports were out of commission because the flare paths, the control towers, and the traffic computers were not functioning. Firemen had to rush emergency mobile generators to hospitals to light operating rooms. Zoos presented special problems. Some animals had to be kept cold, and some hot. Aquariums had to be heated and oxygenated. Keepers wrapped delicate monkeys in shawls, but that was not practical with shivering tigers. The New York Hilton Hotel distributed 30,000 wax candles.

That afternoon the engineer at the console of the Energy

Control Center of Consolidated Edison Company of New York had suddenly seen the needles of his instruments dancing all over the dials. There was nothing he could do. He could not move as fast as the electrons, which were surging backward and forward over 80,000 square miles of the grid area. The object of the grid, or power pool, was to transfer loads from one part of the system to another. Peak loads happened at different times, and the grid was to switch a surplus in one area to relieve a shortage in another at the behest of the master computer. At about 5:15 P.M. New York City was approaching its peak load without any premonitions of difficulties, but up-country, across the border in Toronto, Ontario, the Canadian component of the grid faltered. The effects were immediately felt everywhere from Canada to New York. The current was draining from one area into another in "the cascade effect." People were not throwing switches; switches were throwing themselves. Human observation and reflexes were too slow; only a computer reacting with electron speed could respond. According to the book of rules the computer should have mastered the situation by breaking circuits, preventing generators from becoming overloaded, and keeping the 360,000 volts going over the transmission lines in directions in which they would be least harmful. Instead, the machine itself, like the human brain receiving too many signals from too many nerve centers, had electronic apoplexy.

Energy Makes History

Energy has always made history. The mastery of fire gave man primacy over the animals. The pent-up energy of the longbow gave its bearers a military advantage over the lancer and the swordsman. The gunpowder cannon toppled the feudal system. The coal of the Industrial Revolution stoked the fires of British imperial ambitions—to find raw materials and markets for the steam-powered products of the mass-production factories and to establish coaling stations for merchant ships and their protecting navy. Internal combustion added new dimensions to military strategy—the airplane and the submarine. Nuclear energy vested the two superpowers, the United States and the Soviet Union, with the nuclear deterrent, or the "balance of terror."

Strategic reserves used to be materials stockpiled against the contingencies of war, like the U.S. naval oil reserves. Today, however, peacetime and military considerations cannot, in energy terms, be differentiated. Domestic, industrial, commercial, security, and foreign policy are all bound together. The situation is complicated by the fact that the giant oil companies, wherever their registered home office may be, are multinational in their affiliations and operations. They come to terms with governments—and may be expropriated by governments.

Seven global energy corporations (nicknamed "The Seven Sisters")—British Petroleum, Exxon, Gulf, Mobil, Shell, Texaco, and Chevron (Standard Oil Co. of California)—effectively control the production, research, exploration, and development of oil throughout the world. Despite nationalizations and OPEC (Organization of Petroleum Exporting Countries) the international commercial corporations in the late 1970s still directly controlled nearly half of the world production of oil, most of the refining capacity, two-thirds of the tanker fleet, and most of the major pipelines. Because 300 billion of the proved 500 billion barrels of world oil reserves are in the Middle East, the corporations directed a great deal of their development activities there. Production in that region was cheap (in the 1960s it was $0.16 per barrel as compared with $1.73 in the United States), and the Arab leaders were amenable. Thus the 6% of the world's population that lived in the United States and consumed 30% of the world's energy output continued to burn fuel at bargain rates.

Then, in 1973, a new political force arose. The Arab oil-producing nations dominating OPEC unilaterally increased the price of crude oil. For the first time the oil-exporting countries took the price-fixing function entirely into their own hands. This was, in part, a reaction to the support of Israel by the United States and other countries. OPEC countries regarded themselves as trading their low-priced oil for increasingly expensive industrial products. In a year the price was again increased, from $2.59 per barrel in January 1973 to $10.95 per barrel in January 1974. The price of Arabian oil delivered to the Atlantic coast of the United States shot up from $3.65 per barrel to $12.25 per barrel. Comparable increases were made at European and Japanese ports. OPEC members could claim that the increase in oil reflected the increase in the free market price of gold from the $42.50 per ounce set by the U.S. Treasury. By July 1977 the base price per barrel of crude oil had been set at $12.70, and by April 1979 it reached $14.54. The days of inexpensive oil were over.

The effects were traumatic. The quadrupling of the price of

oil led to an immediate demand for other sources of energy, and their prices soared. Oil-importing countries reacted with oil conservation measures to reduce the effects of the price increase on their balance of payments. Scheduled flights of aircraft were reduced. Ships were detained in ports. Manufacturers of automobiles, aircraft, petrochemicals, man-made textiles, and other oil-dependent operations laid off workers. In the United States, where long lines of motorists formed at filling stations, Pres. Richard M. Nixon announced "Project Independence," an energy policy to enhance national security by making the country essentially self-sufficient in energy by 1980. Great Britain accelerated its exploitation of the North Sea oil fields. Many countries speeded up their nuclear energy development, purchasing many U.S.-designed nuclear power plants and enriched uranium. The increased cost of imported oil had a shattering effect on the U.S. balance of payments, helping to drive down the value of the dollar, while the money markets could not cope with the cascade of "petrodollars" flowing to the oil producers.

The worst victims were those less-developed countries that had got themselves trapped in the oil economy. One of the hopes for the hungry world had been the "miracle grains," high-yielding wheat and rice. High yields, however, meant high inputs of fertilizer, pumped irrigation water, pesticides to protect the growing crops, tractors, and transport. All these depended on oil that these countries could not afford to import.

The big oil corporations were meanwhile diversifying into other sources of energy—coal, the greatest reserve of fossil fuel; shale; geothermal energy; and nuclear fission. They could safely do so because, even if all of their crude oil reserves and production assets were taken over by governments, this would amount to only $32 billion of a total of $134 billion, or less than 25% of the total gross investment of the industry in fixed assets in foreign countries.

What Is an "Energy Crisis"?

As has been seen, from cosmic energy to cell energy and everything in between, there is no shortage of energy. It depends upon how it is to be put to work. If we are hungry, we need food calories. If we are cold, we need heat calories. If we have an automobile, we need gasoline. If we have a television set we need electricity. We can have energy from present photosynthesis (food) or past photosynthesis (fossil fuels);

from sunbeams; from the wind, waves, and waterfalls; and from the crust of the Earth or the heart of the atom.

Oil took over because it was needed as fuel for the internal combustion engine, for heating and power generation because it was fluid and convenient to handle, and, mostly, because it was cheap. Coal was difficult to get and clumsy to handle because it was a solid material. It could be gasified and liquefied, but both processes were expensive. Other energy sources, such as the sun, wind, waves, and geothermal steam, also could not compete in terms of convenience and economy.

The so-called energy crisis that began in 1973 was an oil crisis. The United States could no longer supply its own rapidly increasing needs with domestic production and thus had become hostage to its imports. Such overseas dependence raised issues of national security and balance of payments.

There was (and is) no world shortage of oil. There could be (and will be) an ultimate shortage if the demands go on increasing at an exponential rate for a resource that is unrenewable. Terms of years are given, but they depend on "proven" or "known" or "potential" reserves. "Proven" means just that—that the field has been actually explored, its quality of oil examined, and its extent established. "Known" means "that is where to look," and "potential" means that the characteristic geological formations indicate the likelihood of oil. The estimates of supply therefore vary. The reliable one, proven reserves, changes, of course, with each new strike. From the figure of nearly 69 billion barrels in 1948, the estimate has increased to more than 500 billion barrels in 1978. A general consensus of the amount of recoverable oil, known and prospective, is about 2 trillion barrels. Most of the potential oil will be found on the continental shelf, the offshore extension of coastal states, and on the slopes of the shelf out to water depths of about 600 yards.

The critical factor, however, is the rate at which oil is being discovered and extracted. By the late 1970s this was about 40–45 million barrels a day, as compared with a world demand (excluding Communist countries) of 48–50 million barrels a day. Clearly, the world has to begin managing with less oil. The decade 2010–2020 will constitute a critical period during which oil supplies will begin to fall dramatically. The oil-producing governments are accepting this projection and are planning to conserve and, therefore, to restrict output. This carries price implications. The OPEC increases in 1977, and again in 1979, averaged 10%. Assuming that prices will continue to rise—and, conceivably, at ever-increasing rates—the resulting overall jump in oil prices in the years ahead could drastically affect the world economy.

These price rises will impose three important conditions. First, oil will be used more economically. Second, prices of other energy sources will become competitive with that of oil. Third, countries will not be able to afford the prices and will be forced into energy self-sufficiency.

The proven reserves of coal are put at approximately 700 billion tons, the equivalent of 3 trillion barrels of oil. Potential reserves may be equivalent to another 12 trillion barrels of oil. The geographic distribution of this solid fuel presents transportation difficulties. Its calorific value on the average is, weight for weight, about half that of oil. Massive capital is needed to establish new mines, to provide coal-cutting and coal-handling machinery, and to develop transportation. Miners, once the underground serfs, are becoming the aristocrats of labor, difficult to recruit and expensive to hire. In energy cost accountancy the amount of energy invested in sinking the mines, in the capital-intensive equipment, and in the mechanized extraction and treatment raises the question of the net energy gain. The other problem is the long lead time required to open up new coal seams, particularly those at great depths.

The same problems occur with nuclear power development, which is an evolutionary process starting with the mining and reduction of ore. As has been noted, the production of fusion energy remains some time away. In fission the first generation of reactors, which came into commercial operation in October 1956, was the natural uranium thermal type. The second generation, using enriched uranium, included the advanced gas-cooled reactors, heavy-water reactors, and light-water reactors. The third generation is the plutonium fast-breeder reactor, operating a uranium-plutonium cycle but consuming a much larger proportion of the uranium than in other types and producing as much fuel as it consumes. Excluding a few experimental breeder reactors, by the late 1970s only the second generation had been reached commercially. To get that far had taken twenty years, and it was reckoned that to have breeders in full commercial operation would take another twenty years. This time scale raises the question of the availability and price of uranium to meet the increasing demand of an increasing number of nonbreeder reactors and the sobering problem of the disposal of the radioactive waste.

The hard economics of the nuclear cycle, from natural uranium to plutonium, is illustrated by recalling that for each pound of nuclear fuel burned in a first-generation reactor, electrical energy equivalent to 7,000 tons of oil is produced; a second-generation reactor would produce the equivalent of 12,000 to 16,000 tons; and the third generation (breeder), no less than 800,000 tons. The breeder would consume its own ash, but the result would be plutonium, easy to convert into bombs, a virulent, untreatable poison, and with a radiation persistence lasting 250,000 years.

Environmental Effects

If we accept the present rate of energy consumption on an increasing scale, something must be done about conventional uses and alternative sources. Something must be done about conservation of energy and the conservation of the environment.

The vocal awareness of violations of the environment by technology is comparatively recent, but such violations have been happening for a very long time. The demand for wood for fuel and charcoal for metallurgy ravaged forests. The forests of the south of England became the sparsely wooded Downs. The coal era of the Industrial Revolution blackened the towns with soot, rotted lungs, piled up slag heaps, and poisoned streams with acid. The oil for the internal combustion engine brought smog-producing exhausts, diesel fumes, and jet noise. The nuclear era generated radioactive waste products and a plethora of man-made plutonium.

Apart from the disfigurements and discomforts, the wastes of our energy system can have an adverse effect on that heat engine, the Earth itself, and the living biosphere it sustains. According to United Nations statistics the use of fossil fuel has increased every year since 1860 at a remarkably constant 4.3%. We have taken the carbon that was withdrawn by photosynthesis hundreds of millions of years ago and locked away in the geological vaults as coal and oil and have released it into our contemporary atmosphere. If the known production of carbon dioxide is compared with the observed atmospheric increase, it turns out that roughly one half of the added carbon has remained in the atmosphere.

Where did the other half go? Undoubtedly the oceans are a sink for carbon dioxide, but during this 100-year interval only the upper layers of the oceans, about 100 yards deep, have come into a chemical near-equilibrium with the atmosphere. This has happened because of the extremely slow mixing between the upper and the deep ocean below the thermocline, the oceanic water layer in which the water temperature decreases rapidly with increasing depth. Estimates of the ability of the ocean to continue to take up carbon dioxide, if we continue to produce it at an exponential rate, suggest that the fraction that the ocean can absorb will be less than 5% in the decades ahead. Even if we stop adding carbon dioxide, it would take between 1,000 and 1,500 years for the added amount already in the atmosphere to decay to one-third.

The other absorber of carbon dioxide is the biosphere, consisting of all living matter on the land and in the ocean. Because trees comprise the largest mass of the biosphere, scientists must determine whether the forests of the world are growing more abundantly because of the increased atmospheric carbon available or are being cut down faster than they are growing so that their carbon absorption is being reduced.

The effects of higher concentrations of carbon dioxide are of great importance because of what has been called the "greenhouse effect." The Earth's surface, whether land or water, emits radiation in the infrared band, half of it in the wavelength interval between 9 and 17 micrometers. Carbon dioxide absorbs some of this infrared radiation in the atmosphere, thus reducing the ability of the radiation from the surface to escape into space. This trapped radiation then flows downward and warms the surface of the Earth.

Based on detailed observations, many theoretical calculations of this effect have been made, and they converge on roughly the same conclusion: the present 325 parts per million by volume (ppmv) of carbon dioxide in the Earth's atmosphere will increase to 400 ppmv by the year 2000. This will mean an average increase in the surface warmth of 1.8° F. By 2050 the concentration will increase to 650 ppmv with surface warming of as much as 5.5° F. In the polar regions the increase can be three to five times greater than the global average. The resulting reduced difference between the low and middle latitudes will be significant because it is the temperature difference between the Equator and the poles that drives the atmospheric heat engine. A reduction in this difference will cause a slowing down of the large-scale flow patterns in the air and in the ocean that transport heat toward the poles.

There are five distinct regimes of ice and snow on the Earth: (1) underground permafrost; (2) the winter snow cover that melts in summer; (3) floating sea ice or pack ice, which survives summer in both polar regions; (4) mountain glaciers, which can occur at any latitude; and (5) the massive ice sheets of Greenland and Antarctica, which contain more than one-third of all the fresh water on Earth and have remained more or less intact for millions of years. In the event of a warming up of the polar regions, snow cover on land would be less, glaciers would partially melt, and areas now subject to permafrost would shrink. The big question is what would happen to the ice packs and the ice sheets?

The melting of the sea ice would not affect the sea level (because the volume of the floating ice is equal to the water it displaces). The melting of the glaciers and the ice caps on land, however, would raise the sea level. If all the glaciers melted, the level would rise about 0.6 yards. If all the ice sheets melted, the level would rise about 65 yards. That would be a long-deferred possibility, however, because the warmed-up atmosphere would hold more vapor, which would increase the snowfall over the ice sheets.

A well-articulated awareness of the potential hazards persuaded statesmen of the need to suspend hydrogen-bomb testing in the atmosphere in order to abate the effects of man-made radioactivity. The climatic effects of carbon dioxide, however, are not the concern of any single government or group of governments—even those whose rising standards of living have contributed to the problem. Ignorance has been alerted, but much greater knowledge has to be acquired. This is an environmental, energy-induced problem, the solution of which requires many important decisions at the international level.

All life is energy. Here is energy affecting all life.

BIBLIOGRAPHY

The New Encyclopaedia Britannica (15th Edition)

Propaedia: This one-volume Outline of Knowledge is organized as a ten-part Circle of Learning, enabling the reader to carry out an orderly plan of study in any field. Its Table of Contents—consisting of 10 parts, 42 divisions, and 189 sections—is an easy topical guide to the *Macropaedia*.

Micropaedia: If interested in a particular subject, the reader can locate it in this ten-volume, alphabetically arranged Ready Reference of brief entries and Index to the *Macropaedia*, where subjects are treated at greater length or in broader contexts.

Macropaedia: These nineteen volumes of Knowledge in Depth contain extended treatments of all the fields of human learning. For information on *Understanding Energy*, for example, consult: Accelerators, Particle; Batteries and Fuel Cells; Chemical Reactions; Coal Mining; Coals; Combustion and Flame; Conservation of Natural Resources; Cosmic Rays; Diesel Engine; Earth as a Planet; Electricity; Electric Motor; Electric Power; Electrochemical Reactions; Electromagnetic Radiation; Electron Tube; Energy Sources; Explosives; Gaseous State; Gases, Industrial and Domestic; Gasoline Engine; Gravitation; Heat; Heat Exchanger; Hydraulics, Applications of; Jet Engine; Laser and Maser; Light; Liquid State; Magnetism; Magnetohydrodynamic Devices; Magnets and Electromagnets; Mechanics, Celestial; Mechanics, Fluid; Natural Gas; Nuclear Fission; Nuclear Fusion; Nuclear Reactor; Nucleus, Atomic; Oil Shales; Oxidation-Reduction Reactions; Perpetual Motion; Petroleum; Phase Changes and Equilibria; Photochemical Reactions; Photoelectric Effect; Plasma State; Pneumatic Devices; Pulsar; Pump; Radioactivity; Refrigeration Equipment; Relativity; Rockets and Missile Systems; Semiconductor Devices; Solid State of Matter; Sound; Star; Steam Power; Sun; Thermionic Devices; Thermodynamics, Principles of; Thermoelectric Devices; Tides; Turbine; Universe, Origin and Evolution of; Universe, Structure and Properties of; Uranium Products and Production; Waterwheel; Windmill. For biographical and geographic entries, check individual names.

Other Publications:

Buggey, J. *The Energy Crisis: What Are Our Choices?* Englewood Cliffs, N.J.: Prentice-Hall, 1976.

Considine, Douglas M., ed. *Energy Technology Handbook.* New York: McGraw-Hill, 1977.

Critser, James R., Jr. *Energy Systems: Solar, Wind, Water, Geothermal.* Ashland, Mass.: Lexington Data, 1978.

Gilliland, Martha W. *Energy Analysis: A New Public Policy Tool.* AAAS Selected Symposia, no. 9. Boulder, Colo.: Westview Press, 1978.

McGraw-Hill Encyclopedia of Energy. Edited by Daniel N. Lapedes. New York: McGraw-Hill, 1976.

Morgan, M. Granger, ed. *Energy and Man: Technical and Social Aspects of Energy.* New York: IEE Press, 1975.

Research and Education Association. *Modern Energy Technology: Nuclear, Coal, Petroleum, Solar, Geothermal, Fuel Cells, Oil Shale, Tar Sands, Organic Wastes.* New York: Research and Education Association, 1975.

Thirring, Hans. *Energy for Man: From Windmills to Nuclear Power.* New York: Harper & Row, 1976.

Willrich, Mason, and others. *Energy and World Politics*. New York: Free Press, 1975.

Workshop on Alternative Energy Strategies. *Energy: Global Prospects, 1985–2000*. New York: McGraw-Hill, 1977.

Picture Credits

*Key to abbreviations used to indicate location of pictures on page: t.—top, c.—center, b.—bottom; *—courtesy. Abbreviations are combined to indicate unusual placement.*

Page 8 *Palais de la Découverte -16 H. Canter-Lund—The Natural History Photographic Agency -21 From "Biological Sciences: An Inquiry into Life," 2d ed. (1968); Harcourt Brace Jovanovich, Inc., New York. By permission of the Biological Sciences Curriculum Study. -28 Grant Heilman -29 (t.,c.,b.) *Dr. Winston J. Brill, University of Wisconsin, Madison -33 Richard Wood—Taurus Photos -36, 66, 72, 75 *Hale Observatories -76 *Radio Astronomy Institute, Stanford University -77 *Hale Observatories -80 Audrian Samivel—Rapho/Photo Researchers -82 *NASA -83 (t.,b.) *Hale Observatories -85 (t.) *High Altitude Observatory, National Center for Atmospheric Research -85 (b.) *High Altitude Observatory, National Center for Atmospheric Research; photo, Dr. Gordon A. Newkirk, Jr. -89 *Hale Observatories -91 *Harvard College Observatory -95 *National Radio Astronomy Observatory -96 *Lawrence Berkeley Laboratory -106 *Los Alamos Scientific Laboratory -110 (t.) *Fermi National Accelerator Laboratory -110 (b.) James Quinn -114 *NASA -117 Illustration by Eraldo Carugati -119 *V. P. Hessler -132, 139 (b.), -147 Ken Firestone -153 Erich Hartmann—Magnum -154 David Moore—Black Star -161 Richard Younker -170 "Fairchild Camera and Instrument Corporation -171 *Motorola, Inc., Semiconductor Group, Phoenix, Ariz. -175 *NASA -180 *Atomic Energy Commission -186 *General Dynamics Corp., Electric Boat Division -187 James Pickerell—Black Star -188 Bill Gillette—Stock, Boston -190, 195 *Los Alamos Scientific Laboratory -199 From "A Mirror on Energy" (1977); Lawrence Livermore Laboratory -201 *Plasma Physics Laboratory, Princeton University -203 (t.,b.) *Lawrence Livermore Laboratory -206 Paolo Koch—Photo Researchers -209 Carl Frank—Photo Researchers -211 Dennis Stock—Magnum -214 Stern—Black Star -217 The Bettmann Archive -219 Tomas D. W. Friedmann—Photo Researchers -221 Michelangelo Durazzo—Magnum -226 Chuck Nicklin -230 Peter Menzel—Stock, Boston -232 Paolo Koch—Photo Researchers -234 Mark Antman—Stock, Boston -236 Bernard D. Casey—Taurus Photos -239 (t.) The Bettmann Archive -239 (b.) Peter Arnold -240 Nelson Merrifield—FPG -244 Nicholas DeVore III—Bruce Coleman, Inc. -246 W. E. Ruth—Bruce Coleman, Inc. -249 Georg Gerster—Rapho/Photo Researchers -250 Illustration by Dave Beckes -252 Ken Firestone

Index

a

d

e

f

g

h

i

j

k

l

m

n

O

P

q

r

S

t

u